T0201080

PROFESSIONAL SURVEYORS AND REAL PROPERTY DESCRIPTIONS

PROFESSIONAL SURVEYORS AND REAL PROPERTY DESCRIPTIONS

Composition, Construction, and Comprehension

STEPHEN V. ESTOPINAL
WENDY LATHROP

WILEY

JOHN WILEY & SONS, INC.

Copyright © 2011 by John Wiley & Sons, Inc. All rights reserved.

Published by John Wiley & Sons, Inc., Hoboken, New Jersey.

Published simultaneously in Canada.

For general information about our other products and services, please contact our Customer Care Department within the United States at (800) 762-2974, outside the United States at (317) 572-3993 or fax (317) 572-4002.

Wiley also publishes its books in a variety of electronic formats. Some content that appears in print may not be available in electronic books. For more information about Wiley products, visit our web site at www.wiley.com.

Library of Congress Cataloging-in-Publication Data:

Estopinal, Stephen V. (Stephen Vincent)
    Professional surveyors and real property descriptions : composition, construction, and comprehension / Stephen V. Estopinal, Wendy Lathrop.
        p. cm.
    Includes index.
    ISBN 978-0-470-54259-0 (cloth: acid-free paper); ISBN 978-1-118-08466-3 (ebk);
    ISBN 978-1-118-08467-0 (ebk); ISBN 978-1-118-08468-7 (ebk); ISBN 978-1-118-08584-4 (ebk);
    ISBN 978-1-118-08630-8 (ebk); ISBN 978-1-118-08632-2 (ebk)
    1. Surveying.   2. Surveys—Plotting.   3. Real property.   4. Map reading.   I. Lathrop, Wendy.
II. Title.
    TA545E87 2012
    526.9—dc23

                                                                            2011013565

Printed in the United States of America

SKY10034964_062322

CONTENTS

FOREWORD

Land. Language. Law.

As readers of documents that address every aspect of land—its location, its character, its physical dimensions—surveyors and title examiners and attorneys must all be intimately familiar with the variety of possible interpretations and misinterpretations. Strange and unusual things happen when language is not clear and the law is therefore misapplied.

Throughout our professional careers, we (the authors of this book) have had a wide variety of experiences, often ordinary, sometimes intriguing, sometimes frustrating, related to land descriptions. Our purpose in undertaking this topic is to improve the written record relating to land and to reduce the legal problems associated with poor or unclear descriptions. When the law must be imposed because of problems with the description, everyone suffers, whether monetarily or emotionally. Real property boundary descriptions, land descriptions, must be clear and concise and preserve all the evidence pertaining to boundary location. Often, land descriptions are the only surviving source of information pertaining to a parcel's location. When these documents are missing, incomplete, or ambiguously written, we play a losing game of forensics in locating limited and full ownership interests on, above, or below the earth's surface.

In this volume we examine the land description in the context of time and history to understand how to improve documents we write and how to increase our understanding of those we read. We include history of language, of record keeping, and of surveying, and illustrate the consequences of failure to communicate with court opinions. We hope that the background information provides a lively context to the topic.

PROFESSIONAL SURVEYORS AND REAL PROPERTY DESCRIPTIONS

CHAPTER 1

INTRODUCTION

1.1 PROPERTY

In general, "property" is something that belongs exclusively to someone, whether that "someone" is an individual, a family, a corporation, or other entity, either private or public. As territorial creatures, humans on this continent tend to protect what we consider to be our own property, to keep others from taking it away from us, and to assure ourselves of the entirety of what our property is. Writing descriptions of what we believe we own and the means by which we acquired it is one means of establishing our claims to that property, and this theme will reappear throughout this text.

There two distinct classifications of property, personal, and real, each treated separately and quite differently by our laws.

1.1.1 Personal Property

Generally, if it is movable, property is personal. If property is not land or interests in land, it is personal.

In terms of the law, there are both tangible and intangible forms of personal property. Movable, tangible items such as furniture, merchandise, and livestock fall into the category of "corporeal" personal property, meaning that it has a corpus, a body. However, we can also have intangible personal property. This includes intellectual property: the thoughts in our heads that result in great inventions and the results of our research in the form of reports or other documents. Intangible personal property also encompasses representations of money, such as stocks and bonds.

1.1.2 Real Property

While there are various distinctions within each of the broad realms of personal and real property, we'll be addressing one very narrow category within the latter, focusing on land. This text is about real property and various ways that we describe it.

Real property, in contrast to personal property, is immovable either in fact or by law. It consists of land, buildings and other physical fixtures to the land, along with whatever rights can be exercised in relation to that land either inside or outside of the boundaries of the tract in the form of interests, which we will define in more detail in the following sections.

1.1.3 Ownership

Black's Law Dictionary defines *ownership* as:

> Collection of rights to use and enjoy property, including right to transmit it to others. . . . The entirety of the powers of use and disposal allowed by law . . . The right of one or more persons to possess and use a thing to the exclusion of others.

This description tells us that "ownership" is not synonymous with "possession." Instead, it includes not only "possession" but also the rights to prevent others from having possession and to exclude them from the property. "Ownership" also includes the right to sell or give away the property in a variety of ways, to divide the land, to put it in a will to future heirs, to allow some to enter the land and to prevent others from using or accessing it. Ownership even includes the right to "waste" land by physically destroying it.

This last right is, of course, limited to some extent by the various laws and regulations preventing us from doing harm to others. So while we may not be able to dump toxic materials onto our land because they will leach into the water table and affect others in our community, we can perhaps excavate a deep cavern or remove broad swaths of forest, actions that would prevent others from using the land for construction or other possible future uses.

In general, ownership provides us various powers of free action that are protected by the legal system. The exercise of this "sovereign right" over our property is somewhat limited by local land use, zoning, and subdivision ordinances, but we have the right to appeal for waivers from such regulation. We are also required to act within a period of time defined by state law in order to protect our ownership from claims of ownership or use by others, a period of time known as a statute of limitation, which varies among the states.

1.1.4 Possession

Now to see the other side of the issue, we'll look at *Black's Law Dictionary* to see what it has to say about *possession:*

> ... The law, in general recognizes two kinds of possession: *actual possession* and *constructive possession*. A person who knowingly has direct physical control over a thing, at a given time, is then in actual possession of it. A person who, although not in actual possession, knowingly has both the power and the intention at a given time to exercise dominion or control over a thing, either directly or through another person or persons, is then in constructive possession of it. ... *[Emphasis added]*

We have two basic flavors of possession: physical and legal. These may be the same, or they may differ considerably—and that has been the cause of many a battle between neighbors or long-lost claimants to real property.

Actual possession means that I am actually, physically on the land. Maybe I don't live on it, but I might be farming it, or cutting trees on it, or fencing it in for my cattle to roam. Or maybe I lease it to someone under the claim that I have a right to offer actual possession to someone else. The area that I occupy or use, or give someone else the right to occupy or use, is the area in my actual possession. This says nothing about my right to possession, merely that I do have it.

In contrast, constructive possession gives me the legal right to be on property even though I might not physically occupy the land. A deed gives me constructive possession; it transferred someone else's rights to possess the land to me, even if I never step foot on it. Perhaps I live in the Caribbean full time and never visit the land for which I have a deed in West Virginia. I still have constructive possession of that land through the deed that announces to the world (or to the world that cares to research it) that I have the right to be on that land when I wish without asking permission from anyone except perhaps those to whom I've given the right of actual possession through a lease or other agreement.

The term *possession* comes from the old English *seisin,* sometimes spelled *seizin.* The root of the ancient verbiage makes it clear that it is distinct from ownership, although one with *seisin* may also actually own the land.

1.2   TITLE AND INTERESTS IN REAL PROPERTY

If ownership is a collection of rights in land, *title* is the union of all the elements that make up ownership, a merger of all the rights in and to property. *Black's Law Dictionary* tells us that "Title is the means whereby the owner of lands has the most possession of his property," thereby uniting the concepts of both ownership and possession.

Much of the legal framework by which we use, own, and convey real property comes from European roots, some of it French (as Napoleonic law in Louisiana), some of it Spanish (as in California, Texas, and other areas formerly under Spanish rule), and much of it English. The most common is the English system, and that will be the primary focus of this text. Due to the various historical bases in different parts of the country, surveyors should always research and be familiar with laws in the areas where they practice.

Much of our legal system relating to real property and the language we use when discussing real property arises from English feudal roots. For this reason, that historic background provides useful context for understanding modern treatment of land and land rights.

1.2.1   The Concept of Title

Private ownership of land is a relatively recent concept in the history of humankind. The rise of royalty in Europe brought with it "ownership" of all the conquered land, meaning that the people actually residing on and working the land were merely there at the pleasure of the monarch.

The concept of a monarch or sovereign owning everything crossed the oceans to the New World, and all the explored and settled lands on this continent were claimed in the name of a monarch who never set eyes nor foot on it. This did not prevent kings and queens from granting lands to settle debts—as the king of England so famously did in granting Pennsylvania (literally, "Penn's Woods") to William Penn—or as favors to those who had provided special services or had particularly pleased the monarch. Of course, such grants ignored the fact that there were already people on the land, Native Americans who had no concept of private land ownership and instead treated it as communal property to be kept in stewardship.

American property law and our language related to land are primarily based on the old English feudal system of ownership, which originated during Europe's Middle Ages. This was a method by which the monarch (who claimed ownership of all the land as holder of the crown) controlled all lands throughout the kingdom. Recognizing that it was impossible to control all the land alone, the monarch granted a feud—also called a fiefdom, a fief, a feoff, or a *fee*—to those who swore loyalty to the crown, or to the lords who in turn had sworn their loyalty to the monarch. This feud or fee was the right to possess the land, but not necessarily ownership of it.

The holders or possessors of the land thus granted were called *tenants,* and *tenure* described the terms of their right to hold the land (their *tenement*). Tenure might consist of a number of bushels of corn to be paid annually, or military service, or any other service or payment demanded by the distant owner. The tenant's rights to the land were also called his *estate,* forming the basis for the modern phrase *real estate.*

The modern term *estate* refers to the degree, extent, and nature of interest that an individual has in real property, with *interests* in land being that person's right, claim, legal share, or title in it. It may be that an estate contains less than full title or interest in land, a matter that will be discussed shortly.

Tenancy, or occupation and possession of land, could be of two sorts, free and unfree, with different rights or interests associated with each. *Free tenure* is the modern freehold estate, to which some of the centuries-old elements still apply. *Unfree tenure* is an estate of less than full ownership, having fewer terms of freedom in holding the title than are available to holders of free tenure. Current equivalents are leases and other limited interests in land. With an unfree tenure, the tenant, or possessor of the land, does not have the same rights of selling, dividing, or willing away the land as does a holder of free tenure.

In the feudal system, the tenant of the land was required to swear loyalty (*fealty*) to the grantor, the lord of the land (*landlord*). The ceremony of swearing fealty was called *homage,* an acknowledgment of the limited right to be on the land but not necessarily to own the land; more precisely, tenants had possession rather than ownership, and could not sell it without the lord's consent or pass it on to their heirs after the tenants' death. If a tenant wished to dispose of land to which he had been granted *seisin* (possession), the lord who had granted that possession (the landlord) retained the right of first refusal, called *primer seisin* (first claim of possession), as well as inheritance tax (*relief*) in the form of a year's worth of yield from the land upon the transfer of real property interests to heirs of a deceased tenant.

A lord's dominion over the property (although technically held in trust for the monarch) was at the expense of his responsibility to protect the tenant's rights and to pay ransom (called *aid*) to restore those rights or retrieve land that was unjustly occupied. At the same time, the lord (or the crown) received a "fine for alienation," or a fee for the free and voluntary transfer of land (the current real estate tax), as well as reversionary rights to the land when the freehold tenant had no heir, a situation still called *escheat* from those early days of private land stewardship. In modern times, the state government in which a property lies gains ownership of it by escheat when a deceased landowner has no heirs and no will.

Terms of "unfree" tenancy included *wardship* and *marriage,* meaning guardianship of a deceased tenant's children until age 21 for boys and until 14 or marriage (whichever happened first) for girls. Under freehold estates, guardianship ended upon the heir turning 15, with the guardian making annual reports to the lord about the profits from the land. Wardship created a situation in which minors below the stipulated ages could not control inherited rights to land; the lord had an obligation to pay for the living expenses of his wards but kept all excess revenue. This system also required the lord's permission to marry, otherwise risking loss of any interests in land that would otherwise have been inherited by the tenant's heirs.

While the most common means of acquiring land rights was *tenure by chivalry* or *knight service* (requiring provision of fully equipped knights to serve 40 days of military service annually—an unpredictable any 40 days, and without the possibility of returning home if the 40 days were completed in the midst of battle), other services to the lord or monarch could also qualify. *Serjeanty* (service) tenure required personal service (perhaps arrows or horses for the militia, or meat for the king's palace), while spiritual tenure required provision of regular religious services. *Frankalmoign* (free alms) entailed a general duty to pray for the soul of the land donor without having to provide other religious services. Churches gained much of their vast holdings through providing various divine services to gain spiritual tenure.

In feudal times *socage* (pronounced *soak-idge*) was a land tenure gained in exchange for small and specific services (agricultural or nonmilitary in nature) or a land tenure for payment of rent in money. This made a tenure by socage much more certain and predictable than a tenure by knight service. Of the two original types of socage, the one remaining today is "free and common socage," in which the services supplied in exchange for rights to land are certain, The certainty of the terms of free and common socage is in sharp contrast to the former *villein socage* or *villeinage* in which the services to be provided were not so certain and resulted in an "unfree" tenure that could not be conveyed by the tenant. Eventually, tenures by knight service were converted to free and common socage tenures.

Leaseholds are the most well known modern example of estates of unfree tenure. A lease, meaning any agreement that creates a land-lord/tenant relationship, is a contract for exclusive possession of land for a specified period of time. At the end of the lease, all rights revert to the lessor (the grantor of the lease), the landlord. Our discussion of "less than freehold" and unfree estates as limited estates will provide additional examples.

1.2.2 Fee Simple

When we speak of *fee* in land (formerly the feudal *fief* or *feoff*), we are referring to title, which is the most complete bundle of interests in a tract of land. The term *fee* in and of itself merely notes that interests can be conveyed by a will, but conditions relating to a transfer by any means may be subject to prior specified terms and stipulations.

Simple means that there are no restrictions placed by others on the land—no liens, no mortgages—and so the interests are fully trans-ferrable. As a result, when we speak of *fee simple* title, we mean a title that is free and clear of any restrictions that would prevent the grantee or new owner the right to use and dispose of the land in whatever way he or she wishes.

Absolute means that there have been no restrictions placed on the land by the *grantor,* the one who gave up the land—no rights of rever-sion or future interests exist. The new owner can convey the land to anyone by any means with no conditions attached to that transfer.

Therefore, *fee simple absolute* is the clearest title, subject only to those conditions agreed to or imposed by the new owner of the land, the recipient who is the *grantee* in the transfer transaction.

1.2.3 Limited Title

Title to real property may be qualified in a variety of ways, and in some instances a *limited title* provides less than 100 percent of the full ownership interests in a tract of land. This may be due to shared or joint ownership so that each partner in title has some percentage of ownership and therefore no single partner has full and independent control over the property.

Condominium ownership is a combination of full fee simple title and limited title; the residential or commercial unit in the condominium is fully owned by the person or entity holding the deed, but ownership of the common areas is shared with every other owner in the condominium. This arrangement prevents any single person from single-handedly acting to dispose of the commonly owned property or to affect its use. Each member of the condominium holds a percentage of ownership interests in those common areas, interests that are both limited and protected by the very arrangement of this particular form of ownership.

1.2.3.1 Fee Tail Estates *Fee tail estates* are limited interests first created during the feudal system, intended to keep property in a family line through successive generations. While most jurisdictions have voided statutes addressing this type of estate in order to eliminate it, fee tail estates were originally created by conveying to an individual and "the heirs of his/her body" to prevent property from going to stepchildren or non–family members after death of the grantee. This fixed line of succession could be a fee tail female (inheritable only by female heirs), fee tail male (going to male heirs), or fee tail general (male or female).

1.2.3.2 Determinable Title *Determinable title* is another form of limited interest in real property. Language in deeds conveying determinable title includes phrases such as *so long as, while, during,* or *until.* These terms of limitation provide for automatic expiration of the purchaser's or grantee's fee simple title and reversion of rights on occurrence of a certain event. This reversion returns title to the grantor of the interests (the grantor being the one who granted the deed conveying interests) or that grantor's heirs (as stated in a will), successors (those receiving interests by means of other conveyances from the grantor), or assigns (those outside the will or chain of title to whom the grantor wishes to grant rights). While the grantee of a determinable title may convey his or her determinable interests, later grantees take title subject to the same conditions as established in the original conveyance even though the word *revert* is not necessarily present in any of the later deeds.

1.2.3.3 Defeasible Title *Defeasible title* is a limited interest created by documents that specify a purpose or conditions under which the real property may be used. The main distinction between determinable title and defeasible title is that determinable titles *will* revert when a certain event occurs, ceases to occur, or does not occur, while defeasible title *may* (or may not) revert. As with determinable title, defeasible titles will state the conditions triggering reversion, along with designation of the recipient of the reverting interests (which can be sold separately from the defeasible or determinable interests). Discerning the difference between determinable and defeasible titles can sometimes be tricky, and the context of the documents granting the original rights must be examined carefully in light of the language used at the time of the transaction and contemporaneous conditions.

A case that may help to illuminate the distinctions between forms of title as discerned from the written documents is *United States Trust Company of New York v. The State of New Jersey*.[1] The core of the dispute begins with a deed issued in 1894 to the United States for an area of Monmouth Beach, New Jersey, in exchange for $2,400. The acquisition came about to comply with an 1875 Congressional Act that provided funds to establish "sites for Life-saving or Life-boat Stations, Houses of Refuge, and sites for Pier-head Beacons."

Nearly a century later, the successors to the original 1894 grantors argued that the title had reverted to them because the United States had ceased to use the property as a lifeboat station in 1965. In that year, the United States vacated its use of the property, but permitted the state of New Jersey to use it for the same purposes as the United States had. In 1968, the United States deeded most of the parcel in question to New Jersey for $29,800. Nearly 20 years later, when the litigation began, the value of the beachfront property was well over six figures, and successors to the original grantors sought to regain this valuable site, basing their suit on deed verbiage that mentioned both the 1875 Act and the lifesaving station purpose.

However, the 1894 deed must be read as a whole to determine if in fact it did create a determinable title. The deed itself was of the boilerplate variety, a standardized form used for all such acquisitions in relation to the 1875 act. It did not contain any words limiting the rights acquired by the United States, and merely cited the act that provided the funding and impetus for the purchase, thereby establishing intent. Furthermore, the price paid in 1894 ($2,400) was well more than a nominal fee, or a sum that would merely satisfy the requirement for payment as an element of a valid contract. The court pointed to another transaction in the same time frame using the same boilerplate deed

---

[1] 543 A.2d 457, New Jersey Superior Court, 1988.

for a similar tract in North Carolina for which the United States paid only $100.

Finding no expressed intent for the property to revert to the grantor if the stated purpose ceased, no other language indicating limitations, and a payment of full fair market value, the United States Trust Company of New York was denied its claim of ownership (based on its interpretation of the 1894 deed as conveying only determinable rights), and the court confirmed full unfettered fee simple rights in the State of New Jersey.

1.2.3.4   Life Estate   Yet another form of limited title is a *life estate*. This is a set of interests conveyed to someone for the period of someone's life—whether that of the grantor, the grantee, or some other specified person. Sally can give Cousin Fred a life estate in the old homestead for so long as he lives, which means that anyone to whom Sally conveys the property must honor Cousin Fred's right to be on the land. At the same time, while Cousin Fred can treat the land as if he owns it, he can't do anything to destroy the future interests of Sally's successors and assigns; he can't subdivide and sell off part of the property, and he can't build a 43-story office building on it without Sally's permission. He can, however, lease the land to a gas company provided that the lease does not exceed Cousin Fred's own rights to be on the land either in terms of time or in terms of access to the site.

Sally can give Cousin Fred a life estate for so long as she lives, so that when Sally passes away, Fred's interests cease unless he is named in the will or he purchases the property from those who are named as heirs. Or Sally can give Cousin Fred a life estate for so long as Sally's husband lives, thereby possibly protecting her husband's and children's interests in the land while providing somewhat less assurance to Cousin Fred that he will be able to finish out his days in the house where he spent his childhood.

All of these scenarios are variations of the life estate, and they are all determinable estates. They cease to exist upon the end of a particular person's life, a very specific condition that definitively terminates the interests of the life estate grantee. It should be noted at this point that a life estate, in any of the forms described, is an example of "less than freehold" or "unfree" estate because of the reversion of rights to the grantor (or the grantor's heirs, successors, or assigns) upon termination of the specified period.

As a carryover from feudal days, we still use the terms *dower* and *curtesy,* each originally being a limited title in a spouse's real property. In English law, dower was a one-third interest allowed to a widow in her deceased husband's real estate, in the form of a life estate after his death. At the same time, curtesy was the life estate given to a widower to any real estate owned by his deceased wife, but it was a full life estate interest rather than the mere fraction granted to women. Modern laws have changed both dower and curtesy from life estates to absolute fee interests in a deceased spouse's estate.

1.2.3.5 *Estate for Years (and Variations)* Very similar to the life estate is the *estate for years,* which is granted for a specified and definite period time – whether for a month or for 2,000 years. The time of its termination is known, certain, and definite, no matter its length. In the United States, railroads often received these kinds of "unfree" tenures in land for periods of time probably considered semipermanent at the time of their creation, such as 50 or 99 years. The difficulty with such long tenures is that the parties—or their successors and assigns—often lose track of the need to renew them, and the grantee may actually be continuing use of the land long after the estate for years has expired. This gives rise to another form of interest in property that will be discussed under easements.

Most beneficial to the grantor, and not always so for the grantee, is the *estate at will,* a variation on the estate for years but a tenancy that may be terminated at any time by the lessor (the one who created the estate) or by the lessee (the one who enjoys the limited interest in the property). This is often a month-to-month tenancy.

An *estate at sufferance* is the lowest grade of estate in real property and the lowest form of unfree tenure, held by one who retains possession of land with no title at all, such as a tenant whose lease has expired. This hanger-on becomes a "tenant at sufferance" as long as the landlord/lessor "suffers" or permits him to remain on the property. An estate at sufferance differs from merely trespassing or squatting on property since the original entry was by the owner's permission.

1.2.3.6 *Quitclaims* Finally in our discussion of limited interests, there is the *quitclaim* deed. Such a document merely releases or relinquishes any rights that the grantor may have in the land, but does not state that the grantor actually had those interests in the first place. There is no claim that the title being transferred is valid, no warranty or guarantee in the title to the land supposedly being conveyed. Therefore, if someone with a better (or more legally defensible) claim to title comes along, the holder of a quitclaim deed may not be able to retain the interests contained in his or her deed. Anyone can sell you the Brooklyn Bridge, but only one entity has legal title to it that will actually give you

ultimate true legal ownership of that structure. Thus, only one entity (the true owner) can provide anything other than a quitclaim deed.

1.2.4 Easements

Black's Law Dictionary defines the term *easement* as:

> An *interest* which one person has in the land of another. . . . An interest in land in and over which it is to be enjoyed, and is distinguishable from a *"license"* which merely confers personal privilege to do some act on the land. *[Emphasis added]*

While "interests" were discussed earlier, we have not yet defined *license,* and the distinction between a license and an easement is important. A license provides very specific rights that can be exercised by only a very specific party during a very specific time, and the rights granted by the license can be revoked if the terms of holding those rights are violated.

For example, a generic driver's license grants a single person a right to drive specific kinds of vehicles (generally only certain four-wheeled vehicles, and not big rigs, school buses, or motorcycles), and the license must be renewed on a regular basis. If the holder of a driver's license maneuvers a vehicle improperly enough times to earn numerous tickets and points, that license can be revoked by the state motor vehicle agency that issued it.

In terms of real property, a license may allow a lumber company to enter a tract of land over a certain route to cut certain kinds of trees (perhaps by size or species) in a specified area for a particular period of time, in exchange for a stated payment or perhaps provision of split logs for the landowner's fireplace. If the holder of the license cuts the wrong trees or does not make the proper payment, the licensor can revoke all rights. A license in land differs from a leasehold (the "unfree tenure" mentioned earlier) in that a lease transfers possession while a license merely excuses actions on land in possession of another that without the license would be considered trespass. The license is revocable at the will of the possessor of the land and conveys no interest in the land. When possessors are not the owners of the land, they may grant no licenses harming or lessening the interests of the owners.

Going a step further, the primary difference between a license and an easement is again that the license is subject to termination by the possessor of the land and conveys no real property interest, while an easement is not revocable and does create an interest in land. Otherwise, while it is easy to point to examples of easements, it is not always simple to distinguish easements from the exercise of other rights. Easements can be in the form of the right to use a roadway across someone else's land, or the right to place a pipeline under the land owned by another, or the right to flood an adjoiner's property.

The parties to an easement are the *dominant estate* and the *servient estate*. The dominant estate is the one benefiting by the easement, the one with dominion and ability to exercise the easement rights, and is conveyed "together with" those rights. The servient estate is the one "subject to" the easement, the one burdened by the right of the dominant estate, the one that must allow and not interfere with the exercise of the easement rights.

There are numerous types of easements, the most common being *appurtenant* easements that travel along with the transfer of both the dominant and servient estates, whether or not the easement is mentioned in later deeds. They generally do not terminate until owners of both the dominant and servient estates agree to termination, unless conditions had been established as previously described for determinable and defeasible title. The term *appurtenant* refers to the attachment of the rights to the land.

In contrast to appurtenant easements are *in gross* easements that generally are nontransferable and cease to exist when the easement holder no longer owns or uses the dominant estate, due either to transfer of title or death. These are the personal easements in gross with which we are most familiar. One of the authors of this book received a call from an elderly lady complaining that her new neighbors would not allow her to use her driveway to get to her garage. It turned out that "her driveway" was actually a secondary means of access running over part of the adjoining lot (despite the presence of a separate means of less direct access to her garage that existed completely on her own property) that she had been using for over 40 years. The resulting agreement with the new neighbors was a *personal easement in gross* to my client, allowing her to continue use for as long as she resided in her house, a right not transferrable to her heirs or to anyone to whom she might rent or sell her house. The easement would terminate when she no longer lived there, even if she still owned the house, and was recorded in a deed clearly outlining these conditions, memorializing the intent and purposes of the agreement.

There are also *commercial easements in gross,* which, unlike personal easements in gross, can be transferred from one party to another for the same purposes, although the dominant estate must negotiate with the owner of the servient estate and possibly also seek regulatory permission for such a transfer. The commercial form of an easement in gross terminates when the purpose for the easement terminates, and can be divided into fractional interests to provide percentage ownership to shared holders of the dominant estate in order to guarantee each a right to use a certain amount of an easement in terms of physical space. It also allows joint use, such as pole attachment agreements or sharing a trench (although in this last situation state law may require new negotiation with the owner of the servient estate).

As an example of a commercial easement in gross, Ms. A conveys a gas pipeline easement to Company X, which may convey its entire interest or only a portion of its interest to Company Y. Upon the demise of Companies X and Y or upon their transfer of interests to another company, the easement does not terminate as long as the use for which the easement was granted continues.

Because of the complexities involved with proper use and transfer of easements in gross and the changing burden that may be placed on the servient estate, the majority of jurisdictions rule in favor of appurtenant easements over easements in gross when there are questions as to the form or nature of the easements involved. Deed language must clearly define the intent of the parties in creating either form of easement, the allowable uses, and any conditions under which the easement will terminate or must be renegotiated.

Aside from these two main categories of easements, there are numerous qualifications describing easements. *Affirmative* easements allow the holder of the dominant estate to perform some action on the servient property, such as the right to install a water pipeline. *Negative* easements prevent the servient estate from performing some action that might otherwise be lawful, such as conservation easements that disallow buildings or tree removal in certain areas, or light easements that prevent construction exceeding a certain height that would obstruct natural light from entering windows in a building on an adjoining site. *Secondary* easements are appurtenant to the actual easement, and provide the right to do what is necessary to fully enjoy the primary easement itself, such as the right to maintain it. Even when not expressed, every easement includes such secondary easements, although it is, of course, in the best interests of all involved when the specific rights and limitations associated with an easement are committed to writing in clear and specific language.

A *right-of-way* is generally the right to use the land of another in a particular linear route, and the term is often used interchangeably with the word *easement*. In some contexts and jurisdictions, the term may also apply to the physical strip of land to which title has been granted in fee simple or defeasible fee simple subject to a particular use, such as for a highway or railroad. The determination of whether *right-of-way* is meant as "easement" or as "strip of land" is often gathered from the full context within the documents creating rights-of-way or research into historical practice. This is one of the reasons why clarity of expression and intent are so important when writing descriptions: is a deed granting only a right to use a long linear tract, or is it granting full title in that strip of land?

The public or private status of a right-of-way, when the term is meant as "easement," establishes who may use it. For a public right-of-way, easement rights are given to the public in general and to every individual person to use the right-of-way for the purposes for which it was granted, such as in a public highway right-of-way, without any special permission needed for anyone to utilize it. Every driver of a motor vehicle has the right to drive on a public road right-of-way. Generally, public utilities have the right to occupy a public right-of-way without more permission than is required by state statutes and regulations.

But only certain named persons or entities have the right to use a private right-of-way. The whole world may not use a private way over privately owned land without permission; the risk is prosecution for trespassing. A utility also does not have the right to enter a private right-of-way without permission of the owners of the underlying title in that private way. Because public rights-of-way sometimes are relocated, the land often reverts to private ownership. A utility that is beneath a public way that has been vacated or terminated by some other means must now negotiate with the private owners of the land if it wishes to remain in place. There are often statutory provisions granting utilities time to negotiate or sometimes even the power to exercise eminent domain to condemn a right-of-way so that its facilities can remain in place. Again, the terms of an easement or right-of-way as expressed in the process of creating those rights are essential for determining who may use a right-of-way for what purposes, and in defining what protective measures have been put in place to protect both the dominant and servient estates as conditions change over time.

Covenants are promises between parties that are not in the form of conveyances, but generally have the effect of restricting use of one party's land for the benefit of the other party when recorded in a deed that transfers title to a property. For example, the owner of a development may place a covenant in deeds for the tracts within the subdivision that no garage may be constructed closer to the road than 50 feet from the rear property line, with the intent of maintaining a certain look or upscale ambience in order to maintain high appraisal values. The covenant is a form of contract, and therefore differs from an easement in that breach of a covenant may result in punitive actions and damages, while abusing an easement may result in its termination.

1.2.4.1 Recorded Easements Recordation of a document provides a means by which the world at large can discover the existence of a land transaction, in this instance for the creation of an easement. In written transfers of property including easement rights, those rights are created by *express grant*.

This method is exactly as the words plainly state: the grant of easement rights is expressed in words specifying the intent and extent of the easement. The express grant may be included in a deed for conveyance of full title that includes easement rights to which the property either is subject (servient) or is the beneficiary (dominant, and conveyed together with the easement rights).

Easements by express grant can be created by deed, by will, or by means of some other written document.

1.2.4.2 Unrecorded Easements But sometimes the parties in a real estate transfer don't quite express their intent in the written transaction, although it is clear by the surrounding circumstances that certain easements were intended to be included in the conveyance. In such instances, easements may be created by *implied grant*. Strictly speaking, implied easements are based on the principle that when grantors sell some portion of their land rather than all of it (whether a subdivision of a parcel or one of several adjoining tracts), they grant by implication all the apparent and visible easements that they as the former landowners previously used in order to reasonably enjoy the portion now being conveyed. For this principle to apply, the area now subject to an implied grant must have been used prior to the subdivision of the overall tract as the means of passing between what are now the newly divided parts of that tract.

For example, Marla owns a parcel that she divides into three contiguous tracts, two of them (Tracts A and C) fronting on two different and parallel public roads, while the one in the middle (Tract B) has no direct frontage (see Figure 1.1). She leases out Tracts A and C to others, but lives on Tract B, and regularly crosses over Tract A to access Tract B. Since she owns both Tract A and Tract B, she does not have an easement over Tract A: an owner cannot have an easement in his or her own land, primarily because there is no need for permission from oneself to use one's own land. But when Marla sells Tract B to Dante, he won't

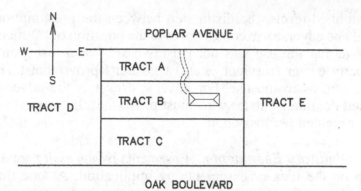

FIGURE 1.1 The dashed line represents the path historically traveled to Poplar Avenue (a public road) from the house on what is now Lot B, creating an implied easement.

have that same privilege to cross Tract A. Because it is a usual and well-established means of accessing Tract B, the right to use that path over Tract A is conveyed as an implied easement to Dante even though it is not specifically mentioned in the deed he obtains from Marla.

If the grantor keeps the land that is subject to the easement, it is an easement by implied grant, but if the grantor keeps the land that requires the easement, this is an *easement by reservation* because the grantor will reserve an easement to himself or herself. But usually this is a matter of strict or absolute necessity to use the property. If another alternative exists, most courts will not create implied reservations.

Because easements by implication can also be created through long, permissive, and equitable use (not overburdening the servient estate), the intention of creating an implied grant must be assessed on a case-by-case basis, looking at the particular facts surrounding the use of the property and the reasonableness of the use of the presumed easement.

Another means of unwritten creation of easement rights is by *presumption of lost grant*. Some jurisdictions presume that if landowners fail to object to use of their land by others, as if an easement existed, then the adverse users must have had a right to be on the property. The presumption of the "lost grant" is that at some point in the past the user of the presumed easement had been given the right to use the property. When no written documents exist, this presumption is particularly applicable to utilities, in the belief that someone must have requested service and the utility complied. Alternatively, for utilities with powers of eminent domain, it is generally assumed that condemnation has occurred and that the time set by the statute of limitations for claiming just compensation has passed.

In simplified terms, the distinction between the presumption of a lost grant and adverse use comes down to the question of "Which party must prove that the use was not permissive?" For a lost grant, it is the property owner (servient estate) who must prove that there was no agreement as to the use. For adverse use, it is the adverse user (presumed dominant estate) who must prove that the property owner had never granted permission.

1.2.4.3 *Statutory Easements*   Easements by *necessity* are a slight variation on the idea of easements by implication. At one time the landlocked parcel was part of a larger tract that had access to public roads. But now there is strict necessity for an easement—the property cannot be used without an access easement. Since division from common ownership of other land having road access creates a new landlocked status, generally the way of necessity is created at the time of the severance, although the specific way did not have to be in use at the time of that division (unlike implied easements). State laws prohibit creation of new landlocked parcels and impose easements by necessity, whether expressed or not, for reasonable enjoyment of tracts, making them usable. The concept of *reasonable* is very fact-specific, and often is the primary source of contention in establishing such easements.

But statutes addressing easements by necessity generally establish a means for a landlocked parcel to access public roads when the grantor has no remaining lands to allow the grantee to cross. In such situations, the owner of the landlocked tract has the force of law supporting a claim of right in negotiating with owners of adjoining lands for an easement providing the landlocked owner access to public roads.

Laws do not require that the shortest and/or most convenient means of access be allowed to the landlocked owner, only that *some* means be available over somebody's land due to necessity. Thus, many adjoining owners can refuse the right to cross as long as one agrees to allow the easement. The ultimate grantors of an easement by necessity over their lands can set the terms of location, width, and other conditions for use (perhaps involving erecting a new gate or locking an existing one).

Another means of creating an easement through the function of law is by *prescription*. This entails long and continuous use of property for a specific purpose in a manner that would allow the owner of the land to know that someone was using his land over a period of time established by state statute as sufficient to allow the owner of the land to evict the trespasser. The use is unrelated to any written document, but is carried out under a claim of right. The rights gained by such use are called *prescriptive rights*. The process of prescription is similar to gaining rights by adverse possession, but reflects only the claim of a right to use the land rather than a claim of ownership.

Once the statutory period of time that would allow the landowner to evict the user of his land has elapsed, the prescriptive easement has been created without any written or recorded document. It is not until both parties acknowledge the situation, either on a friendly basis or under the threat of lawsuit, that the prescriptive rights are set down in writing and recorded in the hall of records to notify all who care to research the title records that the easement does exist.

The most forceful statutory means of creating easements is the exercise of the power of *eminent domain,* or the right of *condemnation.* Lawmakers at both the state and federal levels have identified certain entities that provide essential services to the public and have created statutes to define who may condemn and under what conditions; the prime considerations are public use and necessity. Government agencies may condemn easements when the purpose is to benefit the public. Therefore, a county may condemn land to create a new public road, but it cannot condemn land to create a new restricted-use parking lot just for police vehicles, an area that the public would not be permitted to enter. Each state has its own statutes establishing which utilities are so vital to public welfare that they are granted powers of eminent domain in order to carry out their public services. Therefore, while every state considers water an essential public utility, not every state considers cable television similarly important, so that the first utility may condemn easements while the second possibly may not.

While statutes acknowledge the establishment of easements by necessity, by prescription, and by condemnation, they do not necessarily require any documentation in writing of these statutory means of creating rights to use someone else's land. Instead, particularly in situations of necessity and prescription, the statutes define the conditions under which these means create legal rights, and these laws are the basis for arguments in lawsuits. Unfortunately, all too often a judge will rule that, yes, an easement does exist, but then forgets to order a deed to be written memorializing the location, size, and purpose of the easement so that it can be recorded as prevention of future duplicative litigation. Surveyors engaged in such cases should remind the lawyers involved that recording the outcome in the public records serves to preserve the court ruling.

1.2.4.4 *Distinguishing between Means of Creating Easements*
In the process of describing so many means of creating easements, some of the distinctions may have blurred. The case of *Custom*

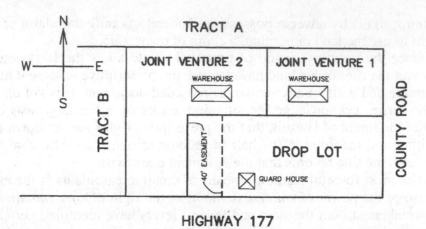

FIGURE 1.2 Custom Warehouse v. Lenertz.

Warehouse v. Lenertz (975 F. Supp. 1240, U.S. District Court for the Eastern District of Missouri, Southeastern Division, 1997) compares several of the methods we have covered and describes how the courts distinguish between them. Figure 1.2 shows the general relationship of the properties involved in this lawsuit.

Attempting to cover all the bases, Custom Warehouse argued it had the right to use an easement over land owned by Frederick Lenertz, Sr., by prescription, by implication, or by necessity. Custom Warehouse wanted the court to declare the "extent and parameters" of the easement it claimed and the location and condition of its deeded ingress/egress easement.

In 1985, Lenertz and his brother had bought about 21 acres from various members of the Howard family. This entire tract was undeveloped farmland, with a creek running through it. Highway 177 bounded it to the south, and a county road formed its eastern boundary.

Lenertz and his brother graded about 12 acres of their acquisition, and covered it with gravel to be a "drop lot" for semi-trailer trucks to pick up and drop off trailers for the nearby Procter & Gamble plant. They installed fencing and a guard shack at the only entrance, from Highway 177, for security purposes. The drop lot began operations in 1985.

Later that same year, Lenertz bought another tract, just north of the first acquisition. At the end of the year, Frederick Lenertz bought out his brother's interest in the northwest part of their drop lot so that he now fully owned that portion in fee by himself.

In 1986, Lenertz bought another 20.59 acres along the southwestern and northwestern edges of the original drop lot, and began grading this along with his 1985 acquisition in order to build a warehouse to use in conjunction with the Procter & Gamble drop lot operations.

A few months later, a joint venture (JV-1) formed to buy Lenertz's land to the north of the drop lot (his 1985 acquisition) to build that warehouse. Lenertz held 40 percent interest in JV-1. Both Lenertz and his brother granted JV-1 a 30-foot-wide easement across the drop lot to provide access to Highway 177 across the land southeast of the JV-1 lands.

A second joint venture (JV-2) formed to buy the property south of JV-1's holdings, to build another warehouse. Lenertz and the two joint ventures (JV-1 and JV-2) executed a "Joint and Mutual Easement Agreement whereby, inter alia, a 40-foot easement was established to provide JV-1 and JV-2 access to Highway 177 along the westernmost border of the Drop Lot property (hereinafter referred to as to the 40-foot easement)." This new 40-foot easement was positioned along the west border of the drop lot, beginning at Highway 177 to the west of the guard shack at the entrance to the drop lot, then running northwest-wardly along that western boundary of the drop lot, and ending at the northwestern boundary of the property near the southwest corner of the warehouse on the IV-2 property.

On the same day that the agreement was made, JV-2 obtained a loan from Jackson Exchange Bank, secured by the property, and Lenertz entered into a lease-purchase option that would allow the other JV-2 partners to buy his interests.

The site began operations in late 1986, with trucks coming in from Highway 177 by the guard house, but then following no prescribed path across the drop lot to get to the warehouses at the north end of the site, varying their routes to go around wherever other trailers were parked. Often, trailers were parked within the 40-foot-wide easement, making it unusable.

In 1992, Jackson Exchange Bank was declared insolvent, and the Federal Deposit Insurance Corporation (FDIC) took over its holdings. JV-2 defaulted on its loan, and the FDIC bought it in 1994. Then JV-1 defaulted on its loan, too, and the FDIC bought its land in 1995. Lenertz continued to operate his other warehouse until August 1995.

In October 1995, the FDIC appraised the two joint venture holdings for a foreclosure auction, to be sold "as is," "where is," and "with all faults." A survey showed only the 40-foot easement along the west edge of the drop lot to benefit the JV-1 and JV-2 lots, in accordance with the deeds. The easement was partly on graded surface and partly on sloping hillside, and the FDIC estimated a cost of $5,000 to improve the easement to make it usable.

Custom Warehouse, Inc., successfully bid on the JV-1 and JV-2 properties, based on the survey, paying a reduced price due to the easement problems. The FDIC had tried to acquire another easement from the Lenertz brothers, but an agreement was never reached (although the Lenertz brothers allowed the FDIC to access the warehouse by means other than the 40-foot easement).

In 1996, Lenertz bought all of his brother's interests in the drop lot, so that he held the entire tract in fee, under the name of FGL Holdings. Custom Warehouse tried to buy a new easement from Lenertz/FGL Holdings, who offered to sell one for $400,000. No deal was reached, but trucks were still allowed to drive in areas other than on the existing 40-foot easement. Lack of a suitable easement was the basis for Custom Warehouse's suit.

The first argument Custom Warehouse raised was for an easement by prescription across the drop lot. But there had not been a single location continuously used for the 10 years required by Missouri law. Further, the use of the drop lot had not been adverse. From 1986 to 1994 JV-1 and JV-2 had owned the warehouses, and Lenertz, as part owner, had permitted trucks to pass from one lot in which he had interest to another. If use begins permissively, it cannot become adverse. The location of a claimed prescriptive easement cannot be "based merely on speculation and conjecture."[2] Nothing was precise, ascertainable, or recognizable about the claimed easement's boundaries, as trucks varied their routes through the drop lot depending on the circumstantial parked location of other vehicles. None of the requirements for a prescriptive easement over the drop lot had been met.

Next, Custom Warehouse argued for an implied easement to cross the drop lot. Implied easements arise when a landowner conveys part of his land to someone else along with a specific passageway over the land that previously had been so open and obvious as to be considered permanent and running with the land (appurtenant).

But, in this case, there had never been any unity of title between the lots involved. In purchasing the lots beyond the drop lot, Lenertz had formed different corporate entities in which he owned different percentages, none being the same ownership he shared with his brother. Lenertz did not have controlling interest in either of the joint ventures or in his partnership with his brother, and he could not create an easement on his own. Lacking unity of title between the joint venture lots and the drop lot, there could be no implied easement.

---

[2]975 F. Supp. 1240 at 1246.

Finally, Custom Warehouse argued for an easement by necessity. Such an easement is created when one party has no means of access to his property from a public road. If one has a legally enforceable right-of-way, he has no right by necessity. Necessity is not a matter of mere convenience. Here, there was an enforceable right-of-way to Highway 177 by virtue of the 40-foot easement created by the Joint and Mutual Easement Agreement. Custom Warehouse argued that the existing easement didn't provide a "reasonable practical" way to access the warehouse or to use the docks in a "counter clockwise fashion." No one disputed that the condition of the existing easement was unsuitable. Instead, the argument was about its position, which would be more reasonable and practical across the drop lot from the southeast corner of the lot. However, as access did exist, no matter how difficult, the court denied any easement by necessity.

1.2.4.5 Estoppels There are times when strict application of laws results in a less than fair outcome. In such instances, the concept of *equity* comes into play, tempering the absolute language of statutes and regulations to carry out the spirit of the law rather than the letter of the law in order to reach a fair or just result. Some states have separate courts for hearing cases of equity, which are either called Chancery Courts or Courts of Equity.

Related to the concept of equity is *estoppel,* which is a legal doctrine preventing one party from misleading another who had a right to rely on the first party's actions, and then taking advantage of the situation to benefit that first party. In other words, if your actions and words lead someone to believe one thing when the situation is different, and relying on your actions leads that other person to harm himself, the doctrine of estoppel will prevent you from the benefit that would have come from the misleading situation. One of the authors of this text had a client who had built a hot tub in what he presumed was his backyard, without the benefit of a survey. The neighbor helped the client build the hot tub, but as soon as it was completed, he notified the client that the hot tub was on the neighbor's property and not on the client's, forbidding the client from trespassing to use it. Because the neighbor knew where the property line was but the client did not, the neighbor's actions purposely misleading the client were frowned on by the court, which invoked the doctrine of estoppel to force the neighbor either to pay the cost of moving the hot tub or to come to an agreement about its use.

Easements can be created by *equitable estoppel,* particularly when no written document can be found to support the origin of the easement, but long continuous use without protest by the servient estate has led the user of the land to believe that a true right exists.

The case of *Spawn v. South Dakota Central Railway Co.* (127 N.W. 648, Supreme Court of South Dakota, 1910) illustrates the application of estoppel when the railroad relied on the actions of the landowner to its own detriment and the landowner's benefit. The railroad had approached Mr. Spawn for permission to build its track across his land. There apparently had been an oral agreement between the parties, because Spawn offered $250 to the railroad if it would build a station near his property within the same public land section so that he could move his agricultural products to market more easily.

The railroad constructed its tracks, built the depot, and began running trains over Spawn's property. But Spawn suddenly decided that he should receive compensation for the use of his land to operate the trains, a detail that was not part of the oral agreements according to the railroad.

From this, we see the importance of written contracts and deeds in preserving evidence of transactions in real property, and laws do exist to enforce such documentation. However, when one party leads another to act in a way that will cause harm to that second party, the separate principles of equity and estoppel provide some protection. In this case, the court noted that the railroad had relied on the oral agreement, had completed all of its obligations under that agreement (building the station), and that Spawn had induced the railroad to cross his land by paying the railroad for that extra obligation. As the landowner, Spawn watched the tracks being constructed over his land, not an overnight process, and had every opportunity to stop construction. But he said and did nothing until after the railroad had made a heavy investment of time, labor, and materials.

While the court noted that writings showing the contractual agreement between the parties would have been more easily enforceable, the actions of both parties showed that the oral agreement had been consummated, and therefore could not be revoked. Its remedy was to decree that an easement in equity over Spawn's land did in fact exist; the means of its creation was equitable estoppel. There had been a clear and definite offer to the railroad for use of the land, reliance on that offer in laying the tracks, and detriment resulting from reliance on the offer, satisfying the three elements of estoppel.

1.2.4.6 Terminating Easements   So far we have discussed only creation of easements. But means of extinguishing or terminating them can also be by written and unwritten methods.

A common means of written termination by local governments is to pass an ordinance *vacating* an easement, and then to memorialize the vacation of the easement in the minutes of the meeting, which are part of the public record.

The dominant and servient estates to an easement may also agree to terminate the easement, extinguishing it by *mutual release*. To remove this easement from use by subsequent parties acquiring the rights of the dominant estate, the best means of termination is to create a written document signed by both parties and then record it in the relevant hall of records.

It is a common misunderstanding that *overuse* (increased burden) or *misuse* of an easement causes it to terminate. This is not entirely true. Imagine that an access easement exists to allow vehicles cross the servient estate to the dominant estate, and the use has always been exercised for two passenger vehicles per day. The owner of the dominant estate decides to tear down the house on her lot and build a warehouse, which will require 40 tractor-trailers to cross the servient estate on a daily basis. The easement does not suddenly terminate due to the increased burden and change in use. But the owner of the servient estate is entitled to an injunction to stop the new increased use while negotiations proceed for increased compensation or for limitation on the hours of truck traffic. If negotiations fail, the court may determine that the easement is being used improperly and terminate it—hopefully, with a decree that will be recorded to prevent any future use of the former access. But the overburdened easement will not terminate until that legal action is finalized.

Obviously, the written approach to extinguishing easements gives rise to fewer questions, if not eliminating them completely, but some situations are so rooted in common sense or common law (the latter meaning well known and understood legal concepts, not from statutes but from court decisions) that no written document is required for termination.

One of these principles is the termination of easements by *merger of title*. If the dominant and servient estates come under the ownership of the same person or entity, the easement will terminate with no action needed. The reason for this is that no one needs his own permission to use his own property. Easements being permissive to someone who does not own the property, they become unnecessary. Should the owner decide to sell either the former dominant estate or the former servient estate, the former easement will need to be created anew, as it extinguished automatically upon merger of title.

It may also happen that the servient estate is destroyed, so that it is impossible for the dominant estate to enjoy its easement rights. For instance, there may be buildings connected by a stairway between them that is located entirely on the servient estate but with easement rights granted to the dominant estate to use those stairs to access the upper floors of the dominant estate's building. If the servient estate burns to the ground, destroying the stairway in the process, the dominant estate is left with no easement due to *destruction of the servient estate*. The dominant estate cannot demand that the owner of the servient property rebuild stairs just to suit the desires of the dominant estate. No written document extinguishing the easement exists in this situation.

Finally, there may be an easement in the records, but it has not been used for a long time. We cannot, however, presume that an easement is *abandoned* merely because the right to use it is not exercised. The courts have stated again and again that nonuse is not abandonment. The dominant estate must have the intent to abandon its rights, and this is made evident by some action making that intent clear. For example, the owner of an access easement over a neighboring property who erects a fence with no gate at the place where the easement exists has made it clear that he does not intend to use the easement.

Abandonment is not accompanied by any written document, and it differs from vacation of an easement, which is accomplished through an official action such as is described above. It can be tricky determining whether abandonment has taken place in some situations, such as when railroads pull up the rails in their easements. Do they intend to repair and reinstall the tracks? How long is a reasonable time to expect that failure to replace the rails indicates abandonment? The answers can be determined only on a case-by-case basis.[3]

1.3 TRANSFERS OF TITLE AND INTERESTS

Whether the interests at stake are for use of real estate or for full title and ownership, there are numerous means of transferring and conveying these interests from one party to another. Once again, some history is useful in understanding the system in place today.

---

[3]The example of rail removal reflects only the physical aspect of railroad abandonment, as the legal process of abandonment is regulated by the federal Surface Transportation Board. However, when the regulatory process of railroad abandonment is complete, there is no written notice that abandonment has occurred.

1.3.1 Written Transfers and Conveyances

The system of record keeping known as a *cadastre,* basically an official register of the value, extent, and ownership of land for the purposes of taxation, began with the Domesday (also sometimes spelled Doomsday) Book. This was a survey of all the lands in England, ordered by William the Conqueror and completed in 1086, with the purpose of determining who held what lands in his kingdom and what taxes should be paid to him as the sovereign. Churches, lords, tenants on the land—all owed the sovereign ruler some form of payment for the privilege of holding the land on which they resided and/or worked.

Patents were written announcements of the regal grant and release of claims, decrees that served as the original deeds from the crown to private landholders. In the United States, we still refer to the original release of a sovereign right to land as a patent, whether from a king, from a federal agency, or from a state. This is the first private ownership of the tract, and it is not uncommon in some areas to have to trace the chain of title or rights of ownership backwards from the current written document all the way to the original patent in order to determine the intent of the parties to that first transaction. It is, after all, what the parties to the transfer of real property wrote down as their intent that must guide all later attempts to locate a tract on the ground or to establish the actual interests conveyed.

Once released to private rather than sovereign ownership, two forms of written transfers could then transfer title and other rights in land to an assigned individual or entity. These two forms survive today, with similar purposes and similar names to their historic predecessors.

The *indenture* is a document transferring rights to land between two or more parties—the minimum of two being the grantor, or one who grants the rights away, and the grantee, who accepts and takes over the rights formerly held by the grantor. There can be multiple grantors and multiple grantees, depending on shared rights. With the indenture, there is a mutual exchange, the grantor transferring real property in exchange for the "consideration" provided by the grantee. Consideration, which provides the motivation or inducement for the transfer to be completed, can be in the form of money, cattle, or a promise for certain actions. When real property transfers took place between family members in certain parts of our country, it was not unusual that "one dollar, love, and consideration" provided adequate indication of a completed contract, serving as the necessary "benefit to one party and detriment to the other."

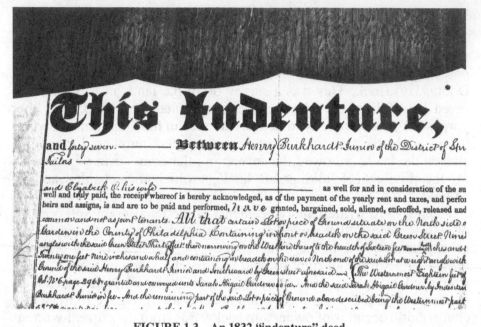

FIGURE 1.3   An 1832 "indenture" deed.

In the days before scanning and copying machines, all documents had to be written by hand, and the indenture document had to be produced in a form that allowed both parties, the grantor and the grantee, to have a copy of the deed upon completion of the real property transaction. It was common to have a desk allowing two scriveners to sit opposite each other at the same desk, writing out the same document simultaneously. If there were multiple parties to the transaction, all of the copies of the document were still written on the same long piece of paper or parchment, which would be cut apart with a wavy or saw-toothed (indented) line upon completion and signing of the duplicate documents (see Figure 1.3). Sometimes a word or phrase would be written between the copies and the cuts made through these writings.

Whether merely indented or also bearing cut writing, if there were ever questions as to the authenticity of documents, the various pieces could be compared to see if they fit together. A perfect fit proved the validity of the claim. Proof of authenticity was particularly important for this kind of transfer, as the grantor promised that the transferred rights were his to convey and that he bore the burden of making the grantee whole if that promise proved to be false, but the grantee was also bound by any conditions set out in the deed. The indenture has evolved into the modern day *warranty deed,* and some more recent examples still open with the language, "Know all ye men by this indenture. . . ."

In contrast to the promises entailed by indentures and warranty deeds was the *deed poll.* In this transfer of real property title and interests, only the party executing it was bound by the deed, and not the grantor or the grantor's successors. There was no promise made that the grantee would be made whole if the grantor turned out not to have any interests to transfer. For this reason, instead of the wavy or indented line allowing copies of the indenture deed to be matched up to prove a claim, a single copy of the deed poll, containing no binding conditions and signed only by the grantor, was cut straight across the top ("polled"). The presence of any conditions would require the grantee also to sign the deed as indication of accepting those conditions, and therefore an indenture would be required.

The modern version of the deed poll is the *sheriff's deed* or the *quitclaim deed.* In such transactions, the grantee takes the risk that the grantor has the right to make the transfer for which the grantee has paid to acquire. Thus, there can be many deeds to sell the Brooklyn Bridge, but only one (from the true owner) will take the form of the indenture deed, while all the others will take the form of a deed poll or quitclaim deed with no guarantee of actual title, interest, or right being conveyed.

Written transfers include *wills,* which are an expression of the deceased's desire or "will" that his or her property will be disposed of in a particular manner. The words of a will are "words of command, and the word 'will' so used is mandatory, comprehensive, and dispositive in nature."[4] The lack of a will can result in default of an estate to the government in an escheat process that automatically occurs when there are no will naming heirs to convey the property and no statutorily defined heirs (such as spouses or immediate family).

Statutory proceedings (meaning fulfillment of processes permitted and outlined by law) can also result in transfers of title or changes in boundaries—presumably written if the judge has remembered to include the appropriate instructions in the final decree. Adjoining landowners disputing the line between them have the legal option to have their argument decided by a court-appointed panel of impartial *boundary line commissioners* or *processioners* (the term varying in different states). The court appointees are then charged with examining evidence of the boundary line, including deeds, plans, testimony, and physical visits to the site, in order to produce a written report. The disputing adjoiners have the option to accept the opinion expressed in

---

[4] *Black's Law Dictionary.*

the report or to turn it down and pursue litigation. The end result of either of these choices should, of course, be a written commemoration of the boundary location to prevent the same arguments over the same line arising between future owners of the same lands.

Multiple owners of a tract may have difficulty dividing it equitably between them, such as when a family of multiple siblings has inherited a large tract of land that varies in quality. Some of it may be suitable for agriculture, some of it may be suitable for timbering, some of it may be suitable for no economically beneficial use. Squabbles over how to divide the land fairly are sometimes settled by dividing the land according to its value and sometimes by establishing size of allotted subdivided portions of the whole. To accomplish either of these outcomes, the multiple owners can petition the court in accordance with the procedures outlined by state statutes, and request *partition* (sometime the executor of a will may make this request). The affected parties can either accept or reject the suggested partition. Whether the report is accepted or the division is finalized through litigation, the judgment establishing the new lines should be formalized in a written description and recorded to prevent future disagreements.

1.3.2 Unwritten Transfers and Conveyances

While a written document is the preferred means of transferring interests or full title in real property, there are recognized means of transfer without that formality. When the interests at stake are simply use of the property rather than full fee title, prescription and estoppel can create easement rights to continue an already established use of someone else's land. Both of these means have been described previously.

Unwritten transfers of title can occur even when no will exists, primarily through rights of survivorship, whether by marriage (dowry and curtesy) or joint ownership arrangements that automatically transfer the interests of the deceased party to the remaining partners.

But when claiming ownership of land without any written support, the claimant often must prove *adverse possession*. While we have mentioned "possession" earlier, we haven't discussed the various forms of possession, "adverse" being just one of many. The most obvious form of possession is *actual possession,* which merely states that someone physically occupies land but says nothing about the right to be there. When actual possession is not based on any written document or identifiable form of inheritance that would have transferred title to the possessor, there is no "color of title," the "color" being even the faintest hint of a right to the possession. In contrast to actual possession is *constructive possession,* meaning that the one claiming ownership has a document supposedly granting the right and title to the real property, whether or not that person actually physically possesses the land.

When actual possession and constructive possession are held by different people with no agreement between them such as a lease or life estate, problems can arise: who has the greatest right? It is then that the claim of adverse possession comes into play. The one with actual possession must show that there was some legally acceptable reason for entering and occupying the land, that it was not merely a matter of squatting on the land with no rights but that there was some ambiguity in the written record giving rise to the belief that the actual possession was appropriate. There is a long list of requirements for proving the validity of an adverse possession claim for title of land in the courts (an action called *quieting title*).

Adverse possession must include the following elements:

- *Open.* It is not hidden or secretive.
- *Notorious.* It is noticed and known in the vicinity.
- *Hostile.* It is without permission and is against the claims of the "true owner."
- *Continuous for the statutory period of time.* Each state sets its own period of time during which the "true owner" has the obligation to eject the adverse possessor (or chain of adverse possessors who tack their time together to perfect a claim) or risk losing rights to the land after the statute of limitations has run.
- *Exclusive.* The adverse claimant fully and exclusively exercises full dominion over the property, preventing the "true owner" from exercising any rights without the adverse claimant's permission.
- *Color of title.* Laws in different states vary as to whether this is a requirement or merely an added condition supporting the validity of the adverse claim.
- *Payment of taxes.* Laws in different states vary as to whether this is a requirement or merely an added show of the validity of the adverse claim. Some states presume that no one would pay taxes on land he didn't own while elsewhere we note that tax collectors are indiscriminate about who pays the tax bill as long as it is paid.

The states differ in opinion as to whether an adverse claim to real property can ripen if it is based on a mistake, but all states bar claims based on fraud and provide protection against the running of the statute of limitations for those under specified disabilities. This protection is generally extended to minors or mental incompetents, but other disabilities sometimes include incarceration or active service in the military. Each state's statutes must be consulted to identify qualifying disabilities within that specific jurisdiction.

Once the adverse claim has been settled in court and the title has been quieted in the formerly adverse claimant (now the new "true owner"), a description of the quieted land must be recorded in order to prevent a repeat of the same litigation. Because an adverse claim is likely to have been for a portion of a property rather than a complete tract, the formal line as established by the court must be clearly described. In such situations, generally some lack of clarity in a prior description gave rise to the problem in the first place.

There is no question as to the ability of government entities and school systems to adversely possess against private owners: they can. But there is a difference of opinion as to whether or not claims of adverse possession can succeed against government entities and schools. Most often, the winning defense of the government or school is that it holds the land for the greater public good, and therefore a member of that public cannot claim against the rest of the public with which he or she shares rights. Exceptions favoring the adverse claimant occur most often when the land is not being held for public use, such as a storage lot for municipal vehicles from which the public is prevented from entering. Additionally, the federal government acknowledged long ago that some of the early surveys of public and government lands were not so accurate, and therefore allowed for owners adjoining public lands held by the Department of the Interior to settle the common boundaries through adverse possession under very specific circumstances. (See 43 US Code §1068, "Lands held in adverse possession; issuance of patent; reservation of minerals; conflicting claims.")

Practical location of a boundary differs from adverse possession in that neither of the parties on each side of the line is sure of the location of the line between them, and the presence of some ambiguity in the language of their deeds (perhaps references to features that no longer exist, or some missing dimensions) prevents documents from solving the question. Over time, the two sides have come to honor a line that both agree is the common boundary, not because they mean to change what is written but merely to settle what they believe is the location of the boundary. This is a practical location, a line to which each side exercises complete dominion as true owner. This may, in fact, effect an unintentional, unwritten change in the boundary, but it can be perfected by a written *boundary line agreement* that fixes and memorializes the practical location for all future parties on both sides. It is not the intent of practical locations and boundary line agreements to alter the written record, but instead to clarify it. Any intentional change in known boundaries must go through the process of subdivision or consolidation as regulated by local land use agencies.

Of course, certain acts of nature can alter boundaries as well. The most well known act increasing land holdings are *accretion* (the gradual and imperceptible increase of land due to natural causes, such as deposits from a river or ocean) and *reliction* (the gradual exposure of land through the retreat of water through natural processes). The actual increased land mass from the process of accretion is *alluvion,* but the terms are often used interchangeably.

Generally, in these gradual natural processes along navigable, tidal, or commercial waterways, boundaries move along with the changes in the water's edge if the body of water was referenced in the deed by which the landowner adjoining the water gained title. Without the stated intent of contact with the water, the ownership may lie with someone else. State statutes address the ownership of alluvion when the water body involved is navigable or tidal in nature. The sovereignty of the states in submerged lands is a relic from English law, when the monarch held them in trust for the public to allow use for navigation as "highways and byways" and for agriculture.

The opposite of the gradual increasing processes of accretion and reliction is *avulsion,* a sudden decrease in land due to erosion of the banks or shoreline. In such instances, the boundaries of a parcel that had been described in reference to the body of water generally freeze at the location immediately prior to the sudden event. Causes may be storms, dam breaks, or changes in stormwater management that reroute water. The question of what is "sudden" and what is "gradual" in order to establish boundaries is often hotly contested, and courts answer it based on the specific facts of a given case.

1.3.3 Statute of Frauds

Eventually, feudalism waned and true private land ownership was possible with more freedom to transfer real property interests. And, eventually, there were disputes about whether those interests had actually been transferred or if there were restrictions and conditions attached to the land. In 1677, the English Parliament passed a law entitled "An Act for Prevention of Frauds and Perjuries," its purpose being to prevent injuries caused by frauds in the absence of a written document spelling out the terms and conditions for a variety of transactions, transfers of real property interests being one of these. The text of Section IV of this statute (29 Charles 2, Chapter 3) reads as follows:

> And be it further enacted by the authority aforesaid that no action shall be brought ... (4) ... upon any contract or sale of lands, tenements or hereditaments, or any interest in or concerning them; (5) or upon any agreement that is not to be performed within the space of one year from the making thereof; (6) unless the agreement upon which such action shall be brought, or some memorandum or note thereof, shall be in writing, and signed by the party to be charged therewith, or some other person thereunto by him lawfully authorized.

The language of this statute presents the requirement for a signed and written document using language we know to be derived from feudalism: tenements and hereditaments. The language clearly addresses any and all interests in real estate, from ownership to any lesser form of possession, except if the agreement for rights in land is to last only a year or less.

While this may seem to say clearly that all transfers of any interest must be in writing (and most of our states have adopted similar statutes), there are, in fact, varying interpretations and applications of the Statute of Frauds in the United States. Some states interpret and apply it to make all oral agreements for interests in real property void, while others states say that the Statute of Frauds merely means that no action (lawsuit) shall be brought before the courts regarding oral agreements. Obviously, this difference of opinion can affect at least the application of estoppel and equity in the different states.

Another aspect of the statute that varies regards the signatures; some states require a signature from only the party who is "charged therewith," referring to whichever party has been assigned an identified responsibility in the document. In some states this is interpreted to mean that only the grantor or lessor must sign, while others deem any real property transaction unenforceable if not signed by both parties.

If we refer back to *Spawn v. South Dakota Central Railway Co.*, it is obvious that having the agreement committed to writing would have avoided the dispute. But the case also provides an example of an exception to the Statute of Frauds. When there is performance (even if only partial) on a presumed contract, those actions can serve as evidence of that agreement. Different judges in different courts in different states may, of course, vary in their opinion of adequate proof of a contract before deciding that one did exist.

1.4 DEEDS

Before the era of general literacy, possession of land was possibly the only evidence of title, with proof of transfers possibly only in the memories of the witnesses to the change in occupation. To memorialize the transfer of physical possession of a freehold estate to a new owner, a symbolic ceremony known as *livery of seisin* involved the feoffor (grantor) giving the feoffee (grantee) some physical thing to represent the land, such as a stick or a handful of earth from the newly acquired real estate, or perhaps a key (the statue in Figure 1.4 commemorates

(a)

FIGURE 1.4a–c This statue of William Penn stands in the rear yard of the county courthouse in New Castle, Delaware. The upper plaque reads: "The citizens of New Castle Delaware presented to WILLIAM PENN (1644–1718) the key of the fort, one turf with a twig upon it, a porringer with river water, and soyle." *[Punctuation added]* The lower plaque is from "The Welcome Society of Pennsylvania" because this area used to be part of Pennsylvania.

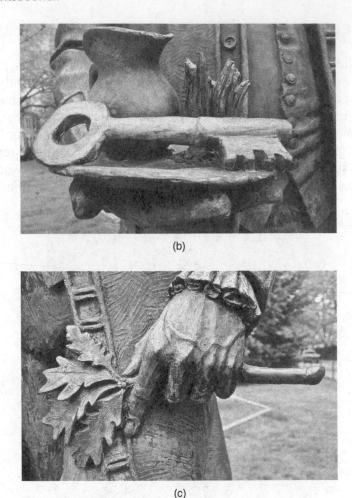

(b)

(c)

FIGURE 1.4   (*Continued*)

such a ceremony). The word *livery* comes from the same root as the word *deliver,* while "seisin" means "possession of the freehold estate." If the ceremony took place on the land, this was termed *livery in deed* ("deed" here referring to "the act" of delivery) and if it took place not on the land but in sight of it, this was termed *livery in law.*

As reading and writing became more commonly held skills, the ceremonial "livery of seisin" began to be accompanied by a written version, a document we term a *deed* (providing evidence of the action or "deed"). Written memorialization was used particularly when the limitations of the estate granted were numerous. The language employed attempted to capture all the intent of the ceremony, severing the stated interests from the grantor and transferring them to the grantee. However, the written deed originally was simply evidence of title and was not the conveyance itself. A writing was not legally required until the enactment of the Statute of Frauds. Eventually, the document replaced the ceremony entirely.

With the historical background we've covered, we can now translate the intent and meaning of more modern documents that continue to use language originating in feudal days. This language can apply to transfers of both full title and partial interests such as easements. Let's look at one example to decipher its intent from the language it uses.

> Know all ye men by this indenture on this fourteenth day of May in the year of One Thousand nine hundred and twenty . . . for consideration of five thousand dollars paid by the party of the second part, that the party of the first part does grant, bargain, sell, alien, enfeoff, release, assign, convey, and confirm unto the party of the second part, his heirs and assigns the hereinafter described property. . . .

We already know from the word *indenture* that this is a warranty deed, one that assures the grantee or recipient that the grantor will stand behind the agreement and contract.

But why are so many other words used in this opening? Couldn't they have been condensed into fewer and more modern terms? In fact, deeds often do employ fewer words these days, but in the process a little of the full and historic impact of the intent has been lost. Each of the words has a unique meaning, a different connotation from the others used in the opening of the deed, and each tells us more about the purpose and effect of the document. (This is not to say that we should continue to employ the full ancient recitation—surely, that language can be replaced with more concise and equally expressive verbiage!)

- *Grant:* Bestow with or without compensation; transfer by deed or writing especially for "incorporeal interests" (having no physical aspects) such as reversionary rights
- *Bargain:* By mutual understanding, contract, agreement, and/or negotiation
- *Sell:* Exchange for valuable consideration
- *Alien:* Transfer to another party (as in alienating oneself from the land)
- *Enfeoff:* Grant possession (as in *fee* or *feoff*)

- *Release:* Give up any right, claim, or privilege
- *Assign:* Transfer for a specific purpose
- *Convey:* Pass or transmit title to another
- *Confirm:* Give assurance, clear away doubts, and approve

Let's look at how these terms interact. If a property is assigned but not aliened, then only an easement has been conveyed; we know that full title is not restricted to specific use of the property and that lack of alienation means that the grantor might retain interests. If a property is granted but no sale is mentioned, we know that there has been no compensation paid, raising questions as to whether full title has passed or merely some lesser interest.

So we see that choice of words in the opening lines of a deed is an important factor in understanding the intent of the parties, every bit as significant as the words chosen to describe the actual land in which interests are being transferred.

Deeds are contracts, and must therefore follow the rules defining the enforceability of any contract. Enforcing the Statute of Frauds consequently prevents some of the problems of oral agreements for transfer of full or partial interests in real property. Depending on the title or interests held by the grantor, the form of the deed will be either a warranty or a quitclaim deed. While the warranty deed provides a guarantee as to the validity of the grantor's rights and interests that he or she is conveying, the quitclaim deed, even lacking that warranty, may in fact convey equally valid title. It all depends on the title held by the grantor, despite the lack of any implication that he or she has good title; the quitclaim deed serves to transfer any interest that the grantor does have at the time that the deed is executed between the grantor and grantee, and that could be nothing at all or it could be full fee simple title.

1.4.1 Legally Sufficient

As contracts, deeds must identify the grantor and the grantee, contain words showing the intent of the parties to transfer the property, and contain a description that is *legally sufficient.* This last phrase means that the deed identifies a property adequately so that it can be uniquely identified (not confused with any other property) and so that a surveyor can locate it on the ground. The courts do not imagine that the general public is able to understand the geometry and references in a deed description.

However, for a surveyor to be able to locate a tract, the description must be clear, complete, and concise. The writer of a description must first determine what the description is intended to accomplish, and then make sure that the words accomplish that intent. Writers of descriptions should not rely on mathematics to establish intent of their documents, but instead should preserve the evidence of the boundaries through references to physical markers and features or other record evidence. More will be said on this matter later.

1.4.2 Abstract of Title

A title search is an examination of records of title to determine whether the presently conveyed title to the property is "good" or if there are defects in the title. The title examiner making the search is looking for anything that would affect the marketability of real property. For instance, the presence of liens (claims against the title in order to satisfy debts) affects marketability in that the satisfaction of the debt may require sale of the property to raise the necessary repayment funds. The existence of liens and other pending litigation (*lis pendens*) mean that title is not "clear." Other conditions affecting marketability of real property are the presence of easements, conditions, restrictions, and covenants; consistency in the record; and lack of access to public roads.

Rather than provide complete copies of all documents relevant to the marketability of a searched property, the title examiner generally provides an *abstract of title,* a condensed version of the history of the land's title. The process of "abstracting" means that the title examiner has reported on only those documents that individual believes to be relevant and significant, and only those portions of those documents the examiner considered important. For this reason, what may be important to a surveyor, who seeks information about the location, configuration, and size of a property, possibly will not appear in the abstract.

Furthermore, the examiner's work includes only a search of public records. There may be privately held records, such as utility location maps, that are not in government offices and therefore not considered public records, but the contents of such records definitely affect interests in property. Surveyors relying on abstracts of title should recognize the limitations of those reports and consider conducting their own additional research.

A further complication for surveyors is that title searching does not always reach back to the origin of a parcel, and so the abstract of title may not provide adequate references for surveying purposes. A scrivener's or typographical error occurring further back than the title search began is likely to go undetected, and consistent re-recording of the same erroneous information can satisfy the title searcher's standards. The number of years that title searchers will go back in the records may be dictated by local standards, or by state statute in those jurisdictions where marketable title acts are in place. Under such circumstances, a state's enactment of a Marketable Title Act required all persons and entities with any interest in real property to file records within a given number of years of that enactment, attempting to create a clean slate by bringing everyone's real property records into public awareness.

1.4.3   Recordation

Recordation refers to the process of creating a public record of transferred real estate interests and ownership. The process of recordation is meant to protect parties acquiring title from other claims to the same interests, allowing anyone to review the written record in order to determine the ownership of land or any claims of rights related to it, that record including deeds, agreements, easements, mortgages, liens, and pending litigation. Other documents found in the public records can include highway plans (for establishment or for cessation/vacation of a state or county road), subdivision and development plats, notices, and wills.

Generally, recordation occurs at the county level for local transfers of real property interests. Requirements regarding acceptance of documents for recording (which may vary between states and even between counties within a given state) can include but are not limited to the following: mandatory contents for recordation; required format for recordation; fee for recordation (including transfer taxes); signature of and acknowledgment by the owner/grantor; approval by the land use agency having authority for subdivisions; and acceptance when dedications are represented on plats.

1.4.3.1   *Notice*   *Black's Law Dictionary* defines *notice* as:

> Information; the result of observation, whether by the senses or the mind; knowledge of the existence of a fact or state of affairs; the means of knowledge. Intelligence by whatever means communicated.

While the Statute of Frauds mandated the existence of some form of writing to convey real property or any interests in it, it was the earlier passage of the *Statute of Enrollments* under King Henry VIII in 1536 that required the grantee of a freehold estate (one gaining full seisin of the property) to record (*enroll*) acquisition of title within six months of execution. Its purpose was to prevent secret transfers of title that could easily be a source of fraud against the grantee.

This statute required deeds for "bargain and sale be made by writing indented sealed and enrolled in one of the King's courts of record at Westminster" or presented to "two justices of the peace, and the clerk of the peace of the same county or counties [where the property was situated]."[5] It came on the heels of the *Statute of Uses*,[6] also passed in 1536, primarily enacted to raise revenues for the king and addressing those who had the use of lands held by another (including in the long list of such interests estates for years, life estates, future in tail interests, and rights of dower and of curtesy). The Statute of Uses acknowledged a legal estate in the one who had the use (and the holder of these interests was taxed accordingly), presumably as a means to prevent fraud.

Until the passage of the Statute of Frauds, writings were not a requirement in the transfer of real estate interests. That meant that public *notice* of the change in owners of real property interests was possible only if the new owner immediately took possession of the land in a manner that was observable and noticeable. Such possession would give *actual notice* to the public, something physically observable. Two basic types of actual notice are *express notice* ("which brings a fact directly home to the party"[7]) and *implied notice,* which provides the means to acquire the knowledge, but not the information itself, such as providing a reference to a deed without quoting the pertinent parts of that deed. *Constructive notice* includes any "information that inquiry would have elicited,"[8] and includes implied actual notice.

There is yet another form, *statutory notice,* provided by the enactment of legislation. Ignorance of the law is not an acceptable defense in the courtroom, and we are all presumed to know the laws guiding our livelihoods because we have statutory notice of them.

1.4.3.2 Racing to Record
It should be remembered that title to real estate does transfer between parties who have contracted through a deed whether or not that document is recorded and filed with a public authority. The purpose of recordation is to establish a uniform approach to settling disputes between various claimants to land within a

---

[5] 27 Henry VIII, Chapter 16, 1536.
[6] 27 Henry VIII, Chapter 10, 1536.
[7] *Black's Law Dictionary*.
[8] *Black's Law Dictionary*.

given jurisdiction, although there is no uniformity between the various states' statutes regarding document recordation. There are, however, four basic categories of recordation acts, all designed to provide notice to the public.

1. **Pure race statutes**

    In some states, the first purchasers of real estate to properly record their deeds will have precedence over all other conveyances from the common source of title (having won the race to the registry of deeds). Because it is possible to record a conveyance even if one has notice of an earlier but unrecorded conveyance, some states employ this approach only for certain conveyances such as mortgages or oil and gas leases in order to preserve equity, although others apply it to all transactions.

2. **Period of grace statutes**

    When transportation and communication were difficult and time consuming, this kind of act was implemented to provide protection to the first recipients of title for a set period of time to allow an adequate chance to record deeds. This is now combined with one of the other forms of recording acts.

3. **Race-notice recording statutes**

    In order to prevail under this kind of act, a later grantee of a parcel must be a *bona fide purchaser* (one who has paid consideration for the full value of the land rather than receiving the property as a gift or inheritance), must record prior to an earlier grantee, and must be without either actual or constructive notice of a prior unrecorded deed or mortgage. (If the earlier grantee records first, this is presumed to give constructive notice to the later grantee.) This kind of statute presumes good faith on the part of the junior grantee as to true lack of notice in the execution of the transaction. The junior grantee of a parcel previously sold to someone else will prevail only if he is unaware of the prior transaction *and* he records his interest prior to the senior grantee.

4. **Pure notice statutes**

    These acts are meant to protect junior grantees against earlier unrecorded conveyances if the junior grantee is a bona fide purchaser *and* has no actual or constructive notice of the prior conveyance (which may or may not have been for full consideration and value). Anyone having notice of unrecorded instruments is barred from claiming better rights based strictly on compared dates of recordation. The difference from race-notice acts is that the junior grantee is protected when the senior grantee records *after* the grant to the junior grantee is executed and *before* the junior grantee records, or even if the junior grantee does not record at all.

To summarize, the rule as to which document is senior is established by a state's statutes. In states where pure race acts are in force, whoever is first to record a document has senior rights. In states where pure notice acts are in force, a subsequent grantee who is a bona fide purchaser without notice of a prior transaction may prevail even if not the first to record. Many states combine these concepts in race-notice acts to overcome some possible inequities that can occur with strict application of one form or the other.

CHAPTER 2

LAND RECORD SYSTEMS

2.1 OVERVIEW

Land record systems are as ancient as the concept of the ownership of land. Territorial control is not exclusively a human phenomenon: all living things require space and resources to, well, live. The means by which living things acquire and maintain this space or control a resource within it varies from simply growing rapidly and crowding out the competition to aggressively asserting control over an area and driving competition away.

Humans operate by the latter means. Prehistoric hunter-gatherer societies, if they were anything like the isolated groups discovered during the age of exploration, exercised a collective ownership of the land. This ownership was protected and expanded by the force of arms. The duration of ownership was a function of the consumption of resources. If game became scarce or the forage wanting, the group moved on. Nomadic people defined their territory by the limits of their physical control as they moved from one region to another.

The adaptation of agriculture increased the duration of possession. As people discovered that resources could be extended and renewable through wise agricultural practices, they developed the land and altered the environment to better suit their needs. The result was that villages, towns, and cities became possible. Collective ownership then extended to fixed areas that were established and defended by the group. The concept of the "state" was born, as an organized and unified community living under a government. The boundaries of a state were usually defined by natural features such as rivers, mountain ranges, forests, deserts, and seas—well-defined and easily identified physical objects. It was not necessary that the limits be precise.

Laws, traditions, and customs developed within states to regulate an individual's activities, responsibilities, and privileges concerning the allocation of resources. As states expanded from mere villages to towns to city-states to kingdoms and finally nations, the land areas involved soon became too large for central control. Family groups or even individuals were granted control of land areas within the state. In Europe, this division took the form of feudal subdivisions, ceding land to the individuals working it in return for loyalty to the central state.

The feudal system, though more efficient than central control, lacked flexibility and efficiency. Feudal landlords discovered that productivity improved if the people they ruled (the land tenants and serfs) were allowed to keep portions of what they produced, and began to assign farm plots to individuals, creating the first individual real property ownership and leading to immense social change. As greater farm yields led to greater prosperity, increasing support for trade and craft specialties in the process, a middle class emerged.

Land ownership within the state could not be left to decision by force of arms. The laws of the state evolved methods of defining and preserving individual real property rights. As individuals began exercising newly increased control over their assigned parcels, personal freedom became a cornerstone of the successful state.

Natural monuments, such as lakes and streams, are often linear. Artificial monuments, such as stones and pipes set in place by humans, are almost exclusively corner markers. Both were used to define the peripheries of new small parcels and the limits of the lands remaining to the lords, identifying the boundaries of each owner's rights. Then, as now, limits of these boundary lines, particularly those dependent on artificial monumentation, *are limited or defined by the location of the property's corners.*

The *precision* or exactness (not to be confused with accuracy) by which boundary lines and corners were defined had to be consistent with land use and value. Furthermore, the parcels needed to be defined in law and reduced to writing to avoid confusion and fraud. And so, even the earliest land descriptions had to identify and apportion the land in terms that a central government could catalog and tax. Land record systems were created so that the enforcer of the law (the government) could be relied upon to insure individual real property possession, use, and enjoyment.

In the United States, there are three land record systems in general use: metes and bounds, the United States Public Lands System, and the platted subdivision. The United States Public Lands System is confined to specific states. The remaining two systems may be encountered in any state.

2.2 METES AND BOUNDS

The oldest of the three land record systems is based on the measurements and the elements defining the perimeters of a parcel, their metes and their bounds. Prior to the introduction of the United States Public Lands System, this was the most common form of describing land in our country, and developed enough momentum in some areas that the newer system was often ignored or reluctantly accepted. One of the authors of this book has worked extensively in jurisdictions where the United States Public Lands System was introduced in the 1800s, yet the ancient and honored metes and bounds system continued in such general use that the referenced section, and even the township and range of many parcels, disappeared from deed documentation.

2.2.1 General History

The metes and bounds land record system developed along with the establishment and enforcement of individual real property rights. Parcels were simply identified by listing or identifying the bounding features or owners along the perimeter, and thus the limits of real property interests. While providing obvious and convenient limits, natural boundaries such as rivers and lakes also often physically divided and prevented efficient use of the land. Unfortunately, natural boundaries did not sufficiently proportion the land for cultivation. As a result, artificial boundaries were created to divide the land into manageable plots; fences kept stock in and others out, and helped in managing land use.

Fallow land—land not subject to grazing or the plow—tended to remain in the control of the sovereign, but eventually, most of the land held by the sovereign body was transferred to individual ownership, at which point it was usually returned to agriculture. During the development of the metes and bounds land record system for individual ownership, the physical limits of an individual claim were therefore obvious: the land was being worked. Parcels were identified by their bounds (boundaries). The land of Smith was bounded by Ford on the north, Wilkes on the south, Fritz on the east, and Browne on the west. A person standing on the land of Smith would have to enter the land of one of the bounding estates when exiting Smith's land. Because all bounds are mutually shared and defined, there are no gaps when land is defined by bounds. For the same reason, there are no overlaps when land is defined by bounds, either.

Mete comes from ancient words meaning "to measure" (*metan* in Old English being just one variant). "Metes" descriptions provide quantified lengths and directions of lines, and often a calculated or estimated area. Such quantification always varies with the sophistication of both the measuring method and the individuals employing that method. Therefore, the size and value of the land, how much is meted out to an individual, are estimates based on guesswork or measurements. The result can be *apparent* overlaps and gaps between tracts that were meant to abut each other and remain distinct from one another.

In a "metes and bounds" description of a real property parcel, combining both of these methods, the "bounds" *define* the limits and the "metes" *estimate* the quantity. The development of the Gunter's chain, the solar compass, electronic distance measuring devices, global positioning systems, and all of the other fantastic advances in measurement science simply serve to refine the quality of the *estimate* of the parcel's shape, dimensions, and quantity. The estimate of a quantity cannot overrule the defined location of a boundary. The existence of gaps or gores (open strips) or overlaps between parcels is not possible when parcels are properly defined by their bounds. If the defined boundary of a parcel is obliterated (cannot be seen) or is lost (cannot be recovered), reliance on only the estimated dimensions (metes) of a parcel can result in apparent gaps or overlaps.

It is the duty of the land surveyor to recover the boundaries of real property parcels. The land surveyor is to identify and expose the physical boundary marks, both natural and artificial, restore that which is obliterated, and find that which is lost. A metes and bounds land record system is anchored on the premise that boundaries are known. A good property description supports that premise.

2.2.2 Legally Sufficient

Because deeds are contracts, they must identify the parties involved and clearly describe the land or interests in land intended to be transferred during the interaction of the grantor and the grantee. Like any other contract, the deed must define who, what, why, when, where, and how much: Who are the grantors and who are the grantees? What is being conveyed (all of a tract, part of a tract, limited or complete interests in a tract, present interests or future interests, and so on)? Why is the deed being executed? (While the flippant answer to this question is "to satisfy the Statute of Frauds that requires any conveyance of land or interests in land to be committed to writing," a better response is "to transfer the desired interests.") When is the transaction taking place? Exactly where on earth is the conveyed tract in which all or part of the grantor's interests are being transferred to the grantee? How much consideration is the grantee paying to acquire the desired interests: fair market value or merely a nominal sum sufficient to show that the transfer is under contract and not a gift?

From this recitation, it may appear that merely answering these questions is legally adequate to transfer land or land interests. The Statute of Frauds requires that all real property transactions:

- Be committed to writing
- Identify the parties involved
- Show the consideration paid by the grantee to the grantor
- Identify the intent of the deed (what did the parties intend as the end result?)
- Describe the conveyance
- Bear the signatures of the parties or their authorized representatives

But deeds must fulfill one additional requirement beyond the scope of general contracts: they must be "legally sufficient." While this may seem redundant, the phrase is critical to satisfying the needs of both parties. A property description is considered legally sufficient if a competent surveyor can locate it on the ground. This means that it must be clear, concise, and complete. The author of a description must first determine what the deed is intended to accomplish, and then assure that the chosen words achieve that intent. It is, after all, the words that identify intent, and not the mathematical closure of the description that establishes it.

While context helps to discern the intent of a deed and description, relying only on context and contemporaneous circumstances can make life difficult for anyone who has to do the additional research to determine that intent, and essentially to determine if the description is legally sufficient. However, there are times when what appears to be vague may in fact be complete enough to legally suffice in locating a property. An example of legal sufficiency when the deed is coupled with local knowledge is illustrated in the case of *Crawley v. Hathaway*.[1]

---

[1]721 N.E.2d 1208, Appellate Court of Illinois, 4th District, 1999.

Purchaser Crawley sued landowner Hathaway to complete the 1995 contract of sale that they had entered into several years before, but Hathaway claimed that the Statute of Frauds would prevent him from honoring the planned conveyance. There were ulterior motives behind the would-be seller's claim.

The document, in its entirety and in the following format, read as follows:

Agreement to Buy 100 Acres More or less, 83 acres of pasture & timber and 19 acres of tillable ground For $90,000 Seller Mark Hathaway Buyer Doug Crawley

Crawley gave Hathaway a deposit check for $7,500 (which Hathaway cashed), and then proceeded to secure financing for the rest of his purchase, which he believed to be about 100 acres of woods and tillable land in pasture and hay, although he was not sure of the specific acreage. Crawley then hired a surveyor to survey the land and prepare a written description. Both Crawley and Hathaway were present to point out the boundaries and assist in determining a point of beginning for the survey and description.

But within four months of the survey's completion, Hathaway changed his mind about the sale, since it seemed more land was being transferred than he wanted to sell to Crawley. So in January of 1995 he refused to complete the sale and instead listed the land with a realtor for $150,000, advertising it as 127 acres, more or less. Crawley sued to force the sale according to their original agreement and description.

After citing the Illinois Statute of Frauds prohibiting actions relating to contracts for the sale of lands unless committed to writing signed by the interested parties, the court stated:

The memorandum is sufficient to satisfy the Statute of Frauds if it contains upon its face the names of the vendor and the vendee, *a description of the property sufficiently definite to identify the same as the subject matter of the contract*, the price, terms and conditions of the sale, and the signature of the party to be charged.[2] *[Emphasis added]*

Both Hathaway and Crawley had met with the surveyor at the time the survey was initiated, and so the survey itself could be considered when determining whether the memorandum was a valid contract. While on site, Hathaway identified for the surveyor the point at which

---

[2]721 N.E.2d 1208 at 1210, citing *Callaghan v. Miller,* 162 N.E.2d 422 at 424.

the survey was to begin. He told the surveyor to follow the road to the north, then follow the road to the east to the tracks, follow the tracks on to the north boundary, and then complete the survey by following the tree line. Hathaway admitted it was his intention to sell to Crawley the wooded area and 19 acres of pasture encompassed by the survey.

From this evidence and the resulting survey, the court could reasonably conclude that both parties knew exactly what property was meant to be conveyed, even though Hathaway was unaware of how many acres it contained. If there was no other parcel of land that could be the subject of the document, the description was clear—and Hathaway owned no other land in the area. Furthermore, parol evidence (verbal testimony in court) was admissible to clarify the site's location, although such evidence cannot supply missing information to contradict or change what is clear. Hathaway himself provided that clarifying parol evidence.

The dissenting judge protested that the description contained "nary a clue as to where the property might be located." The survey does not call for the contract and the contract does not refer to the survey. He also pointed out that the contract listed Hathaway as the seller, but not as the owner, and these could be two separate people, as when an agent sells on behalf of an owner.

And so we see that what qualifies as "legally sufficient" must sometimes be defined by circumstances beyond the actual written property description. It is our intent to provide enough guidance in this text so that litigation like *Crawley v. Hathaway* won't be necessary to detect the meaning of descriptions we write or to establish the location, shape, and size of the lands we describe.

2.3   UNITED STATES PUBLIC LANDS SYSTEM

The use of the United States Public Lands System is limited to specific States. Survey instrumentation, survey methods, corner monumentation, and archival responsibilities under this system vary from state to state and even within a state. The general description we present here is not intended to be a detailed or comprehensive analysis of the system.

2.3.1   History

At the close of the 18th century, the newly established federal government of the United States lacked funding. Raising taxes is a difficult assignment for a democratically elected government, and the United

States had just been created by a rebellion against taxation, so levying taxes to fund the new government was doubly difficult. But the new government did have land—millions of square miles of it. In 1784, the Congress assigned a committee, chaired by Thomas Jefferson, to develop a method by which the land could be sold to the flood of immigrants that jammed the East Coast and flowed inland.

The old European metes and bounds system was not up to the task of creating hundreds of thousands parcels of land that were small enough for a family farm, described in terms that were easily understood, and capable of further division by illiterate land owners. A new system had to be created to accomplish all of these objectives for vendors (the government and its agents) and vendees (the settlers) who often had never seen the land being conveyed.

So Congress passed the Land Ordinance of 1785 to establish the system we currently called the Public Land Survey System as a methodical means of surveying, dividing, and disposing of the lands held in trust for the public by the federal government. Establishing the direct sale of land by the government to a private purchaser meant a change from older practices and their resultant abuses, with Congress acting in the place of the former royal sovereignty in disposing of the public domain for the common good.

A rectangular system was thought to be the most logical means of dividing lands, although perhaps not the most equitable due to variations in land quality between the various rectangles. Regardless of quality, the nation's public lands were divided first into townships (eventually but not originally a nominal six miles on each side), then into sections intended to measure one mile by one mile, then into quarter sections and smaller regular fractional parts. The ideal (but not always realistic) township and its perfectly square sections are shown in Figure 2.1.

The first attempts at uniform surveying of the public lands had varying results, depending on the skills and knowledge of the surveyors gathered by Thomas Hutchins in 1785, the geographer in charge of establishing the first seven ranges upon which the national system would be based and from which it was intended to be expanded.

The full and colorful history of early attempts to survey wild areas prior to federal sale of lands is beyond the scope of this text, but over time one point is clear: the parcel corner became the foundation of the United States Public Lands System (USPLS). As with the metes and bounds land record system, the *actual locations of the corners define the boundaries*. In the USPLS these corners are established by the United States government surveyor, and their locations are immutable.

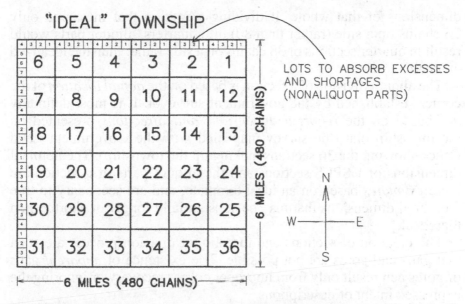

"IDEAL" TOWNSHIP

LOTS TO ABSORB EXCESSES AND SHORTAGES (NONALIQUOT PARTS)

FIGURE 2.1 The ideal township was the template for divisions of public lands. In reality, not one township in the United States has the dimensions portrayed in the ideal. The fact that the Earth is a spheroid, along with difficulties of terrain, conditions, and the large measurement errors inherent in the instruments employed, cause the ideal township to be an impossibility.

2.3.2 Aliquot Division

The USPLS land record system provides for a specific and regimented method of subsequent division of a section, with a *section* defined as being bounded by lines drawn between the corners as established by the government land surveyor. The ideal section is presented as a perfect square of 80 chains (one mile) on a side. But, in reality, few if any of the millions of section corners set by the government surveyors are perfectly 80 chains apart, and it is a rare section that is a perfect square. In cases where the government surveyor established corners at the midpoint between the section corners, these midpoints being called quarter corners, none were physically at the perfect midpoint.

The division of a section or subdivisions of sections into regular fractional parts without any remaining or leftover parts is called *aliquot division,* with each of the resulting pieces called *aliquot parts*. This process is based exclusively on the actual location of the section corners set by the government surveyor. The sum of the aliquot parts equals the whole of the original tract of land and disregards the nominal, idealized dimensions of that whole. If dividing a section that measured only 76 chains on a side (rather than 80) into quarters, aliquot parts would result in quarter sections of 38 chains on a side rather than the idealized 40 chains.

The aliquot division of a section is based on the *actual location* of the corners established by the government surveyor. It is most definitely *not* based on the *reported dimensions and directions* presented on the township plat (the survey map filed with the Government Land Office showing the 36 sections making up the township). The nominal dimensions of USPLS sections and the aliquot parts of a section are mere *estimates* based on an ideal template, and are secondary to true measured dimensions that may be substantially longer or shorter than that ideal.

The creation of sections and the aliquot division process are such that gaps and gores are not possible. The existence of *apparent* gaps or gores can result only from improper corner recovery or ignoring the expressed intent of descriptions.

2.3.2.1 Irregular and Fractional Sections The USPLS land record system is based on an idealized flat world that can easily be divided into regular rectangular parts. But the convergence of north/south lines as they stretch toward the poles causes the shortening of east/west lines, warping the rectangles and lessening the area they contain. To maintain the largest possible number of regular sections within a township, all compensation for the effect of a round earth on straight lines is thrown into the sections along the north and west boundaries within a given township. This means that sections 1 through 7, 18, 19, 30 and 31 (refer back to Figure 2.1) are all designated as fractional sections that will never be perfect squares and will never contain the full 640 acres of a nominal regular section.

The creators of the USPLS also recognized and provided for situations where significant natural features would disrupt any attempt to conform to the ideal section, and so irregular or fractional sections occur at the borders of large lakes, bays, and other water bodies. Fractional sections are not divided using aliquot system guidelines.

2.3.2.2 USPLS Lots In the USPLS, a *lot* is the smallest area of land that is suitable for building or occupation, and is normally created by platting. These are parcels that cannot be described as aliquot parts of a section.

While lots generally lie within a given section, *tracts* encompass area within more than one section, generally within a single township but possibly including land in others. Lots exist along the north and west boundaries of a section to accommodate the excess or deficiency in measurement of that section (see Figure 2.2). Lots were also created along riparian boundaries of fractional sections.

Aside from the usual irregular measurements caused by convergence, as discussed earlier, the USPLS provides for the creation of lots instead of aliquot division in fractional sections fronting on navigable waterways or other interruptions of the regular section pattern. Refer to the United States Bureau of Land Management's "Manual of Surveying Instructions" for detailed procedures and policies of the USPLS divisions.

2.3.2.3   Nonaliquot Division   If a division of a section is performed that does not conform to the guidelines provided for in the USPLS land record system, that division is not an aliquot division. Any parcel so created would then be controlled by the metes and bounds land record system or the platted subdivision land record system. The distinction is very important, for the rules directing reestablishment of lost corners in the USPLS will not apply to nonaliquot division.

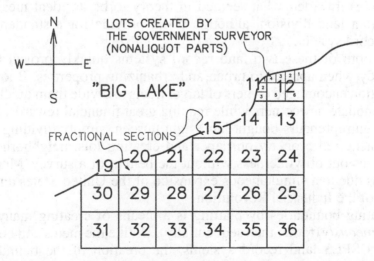

FIGURE 2.2   **The lots created along the boundary of fractional sections were rarely monumented. The Bureau of Land Management's** *Manual of Instructions* **does not provide for aliquot division of such sections and lots.**

2.4 PLATTED SUBDIVISIONS

Although widely used and often mandated by regulatory authorities, platted subdivisions have an Achilles heel: the system is dependent on the survival of the recorded subdivision plat. Descriptions that rely solely upon reference to a recorded plat will be rendered obscure should that document be lost or destroyed.

2.4.1 History

In the metes and bounds system, smaller tracts of land are divided from larger tracts one at a time. Transfers of these lots are *sequential conveyances,* creating *junior and senior parcels,* as newer junior lots are divided out of and conveyed from older, more senior parcels. In the instance of any deficiency or surplus in the original tract being divided (as compared to stated dimensions or acreage), the last lot created (the most junior tract) receives whatever is left after the prior sequence of more senior conveyances. The grantor can convey only what he has and no more. This means that the most junior lot may be smaller or larger or differently shaped than expected or described.

In contrast, the USPLS was meant to be a methodical approach to create regularly shaped, equal areas out of the vast public lands held by the federal government. Actual topography (and less-than-perfect surveying) thwarted what seemed in theory to be an ideal means of egalitarian land division, although it did provide for a standardized land record system.

But both of these two land record systems quickly proved to be unwieldy when applied to urban and urbanizing properties. Booming population encouraged owners of larger tracts to divide them quickly to accommodate newcomers while reaping great financial rewards. Elsewhere, entrepreneurs bought land sight unseen, simply dividing their newly acquired property on paper and creating lots, neighborhoods, and towns out of wilderness without the benefit of a survey. Much of this was due to a simultaneous expansion of the United States and the arrival of the Industrial Revolution.

Creating boundaries by platting is a means of creating many lots *simultaneously* (rather than sequentially). Unlike the metes and bounds or the USPLS land record systems, the creation of the boundaries of lots in this manner often precedes the monumentation of the corners.

Once the corners of the subdivision lots are monumented and relied upon by the landowners, the boundaries are defined by the location of the corners. Within a given platted subdivision, gaps, gores, and overlaps are not possible, as all lots are created simultaneously without junior or senior ranking. *Apparent* gaps and gores can exist only if there has been a misinterpretation of the field evidence. But there can be gaps, gores, and overlaps between separate platted subdivisions, which also can have junior and senior rights between them.

2.4.2 Recorded Plats

Public recordation of a subdivision plat means that the document is given an additional layer of credibility and authority. Storage with the proper authority (county recorder, for example) prevents alterations and allows the world to reference and rely on the plat. Documents kept in private records are not always accorded the same legal respect, and a user of privately held documents may have to prove that these records are above suspicion, even though their quality may far surpass those kept publicly. Recordation informs us about the authority of a plat, but does not attest to its quality.

Areas shown on plats as intended for public dedication, whether as streets, alleys, town squares, or parks, may not represent actual completed dedication. There must be acceptance of the dedication before the public nature of the dedication is perfected; both the offer and the acceptance of that offer are necessary elements in accomplishing the dedication. Sometimes the acceptance appears on the face of the plat, but sometimes a local ordinance finalizes and memorializes the dedication. However, the offer for dedication does not expire until officially turned down by the would-be accepter, even if many years after the original offer.

Even when streets and parks have been dedicated, actual construction of these improvements may never materialize, resulting in "paper streets" and other invisible features. Yet until a formal vacation of these dedicated areas occurs, or there is an official refusal to accept them, those purchasing property based on reference to the recorded plat have every right to rely on the features shown on it as part of their purchase contract.

Obviously, neither physical lack of improvement nor later vacation and closing of depicted public areas appear on the recorded subdivision plat, and the description writer has a responsibility to assure that the document he or she drafts reflects *current* conditions rather than those frozen in time when the plat was recorded. Aside from physically nonexistent features, there is also the possibility of more recently added roads and public areas, or lot consolidations and resubdivisions.

A question of ownership also arises when the sole description is by reference to a plat: does ownership run to the sideline or centerline of the streets shown? In some localities, all streets, alleys, and highways are owned by the abutting landowners, whether or not dedication has occurred, as the government accepts only an easement for the benefit of the public. Elsewhere, roads are privately owned to the centerline because there is no intent to dedicate. In still other places, the government takes fee ownership to the full right-of-way width of public roads.

2.4.3 State and Local Regulations

Jurisdictions may require dedication statements with signatures of the grantors prior to acceptance of the offer. These statements sometimes appear on the face of the survey or subdivision plat, but sometimes appear only in local records, and may not even appear in the county hall of records. To facilitate full recovery of necessary information, check with each locality for the procedures in place. Requirements may also vary between state and federal agencies involved in land transactions.

A reference to a subdivision plat, or to a tax block and tax lot may or may not be an adequate description for the purposes of conveying real property interests. The statutes in each state must be checked for the acceptability of such a recitation in place of another form of description. It is also possible that conditions may change after the recordation of a subdivision plat, with some lots redivided or consolidated after approval and filing of the plat. If a reference to a subdivision or other filed plat is acceptable in a given jurisdiction, include the full name of the plan, the date recorded, and the plat number or other recordation information.

A reference to a platted subdivision is not the same as a reference to tax map parcel information, which can change whenever the tax assessor or maintainer of assessor's maps changes the reference system or otherwise corrects or alters the tax assessment map. A reference to a tax map incorporates any plotting errors that are present, and includes neither metes nor bounds of the subject of the land transaction. While an unadorned reference to a tax parcel may be a legally acceptable property description in some places, it provides no evidence of the tract's boundaries or history, and provides only the most minimal secondhand information.

Regulations establish the contents of a platted subdivision, such as lot lines and their dimensions and directions, monumentation, possibly deed references, and other useful evidence of the subdivision's interior and exterior boundaries. Jurisdictional regulations also establish how much detail must be included for curved lines (the most common being radius, arc length, central angle or "delta," chord bearing, and chord length) as well as establishing the direction in which straight lines must be described (clockwise, for example).

2.4.4   Federal, State, and Local Government Maps

An increasing proliferation of public maps is available online. Readily available local government maps can include assessor's maps (as well as the tax rolls identifying owners), civic planning and zoning maps, demographics, and an endless number of other graphic representations. A wide range of state and federal documents include flood hazard mapping, topographic maps, mineral maps, wetland maps, and many more.

Many of these maps include boundary information of varying quality and accuracy. United States Geological Survey (USGS) maps, for example, show the *approximate* location of township, range, and section lines. These presentations are not the recovered or proven location of USPLS sections and should not be relied upon as defining real property boundaries.

Maps and plats that are produced for one purpose, such as topographic mapping or municipal zoning, should never be used for a purpose that the map developers did not intend, such as real property boundary determinations from the USGS depiction of township, range, and section lines.

2.4.5   Linear Tracts

When a property is linear in form, it cannot be described by aliquot parts, but it can be described by metes and bounds. Such a description includes the bearings, distances, and bounding references of each of the tract's sides. But such a tract can also be defined by a reference line, with a width on each side of that reference line. This type of description is common in defining roads, railroads, utility rights-of-way, and other long strips of land that would be cumbersome to describe in another manner.

2.4.5.1   Defining Control   The most practical control or reference foundation for particularly linear tracts is usually linear as well. The anticipated use of the control will determine its identity and nature.

2.4.5.1.1   Centerline   When the width of the linear tract is the same on each side of the reference line, that reference line is the centerline, and can be referred to as such. It is important that the widths be stated as "on *each* side of the centerline" rather than "on *either side* of the centerline," as is commonly written. The first phrasing tells us that the width of the right-of-way is the same on both sides of the reference line, but the second version sounds like it offers a choice between one side or the other, either this side or that one. For absolute clarity, choose words with the fewest possible interpretations and the most specific definitions.

Over time, however, it is possible that a linear tract could be widened on one side and not the other, or subtracted from on one side of that reference line. As an example, an old colonial road might have started as a two-rod road, 33 feet wide, centered on a line that was established and surveyed by the road viewers appointed by the county judge in 1763. When created, the road could be located by a description of its centerline and its stated width.

While that two-rod road would have been wide enough to be considered a major highway at the time of its creation, as time passed and the population increased, use of the road by more vehicles traveling at faster rates made it inadequate and unsafe. Standards for public roads (as opposed to private roads) now require road widening when development will increase traffic along the roads on which the development fronts. Local and county land use ordinances often require any development that will increase traffic to contribute to safe passage along the roads by allocating an additional strip of land along the existing road right-of-way. The result is that the original centerline of the original road is no longer in the center of the newly widened roadway (unless equal widening coincidentally occurs on both sides of the centerline at the same time).

Examples of other reasons for variable widths include local government land acquisitions for bridges, drainage, or slope maintenance, creating "bump outs" of a different width than the rest of the linear tract. Figure 2.3 shows a variety of changes in road right-of-way width, all related to the same reference line that was the original centerline or baseline.

When describing a linear tract by its centerline (whether or not it is of equal width on each side of its centerline), the writer must be aware of the angle at which the described centerline intersects with and terminates at the boundaries of the tract through which it crosses. Unless that intersection is either perpendicular or radial to the centerline, there will be slivers of land either included when they should be excluded, or excluded when they should be included. Figure 2.4 is an illustration of the problem. In such instances, the limiting bounds must be explicitly noted, with the linear tract extending to and limited by those bounds.

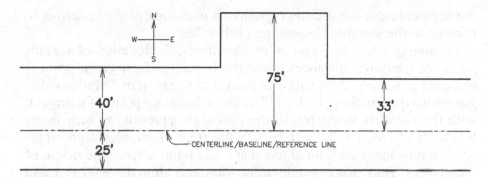

FIGURE 2.3 The original centerline of the original road may no longer be the center of the current right-of-way, as widening occurs at different times at different widths for different purposes. However, the ability to recreate the original centerline may be the key to determining current property rights based on ancient right-of-way locations and widths.

2.4.5.1.2 Baseline
Long strips of land used for highways and railroads are related to a line often referred to as the *centerline,* whether or not that line is still physically an equal distance from each of the sidelines of the right-of-way. A more appropriate name for such an offset reference line is *baseline.* Stationing becomes important in defining

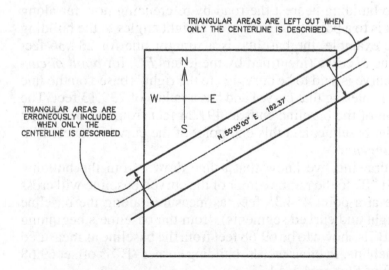

"..BEING A 50' WIDE RIGHT—OF—WAY CENTERED ON A LINE RUNNING N 55° 35' 00" E .."

FIGURE 2.4 While describing the centerline of a right-of-way that has an equal width on each side of that centerline may seem the most efficient approach, it can lead to the erroneous inclusion or exclusion of lands. While a centerline description can be employed, use additional clauses to clarify the intent at the intersection of the centerline with tract lines that are neither perpendicular nor radial to the described centerline.

the segments that are different widths on each side of the baseline, in relation to the identified beginning of that line.

Stationing refers to a system of identifying the location of various points at measured distances along the baseline. Each given point is assigned a "station" designation in this reference system. "Stations" begin with a designation of "0 + 00" at the originating point of a project, with the numbers to the left of the plus sign representing how many hundreds of feet and the numbers to the right of the plus sign representing how many additional feet a specific point is from the origin of the project. Thus, for example, a location at Station 12 + 45 is 1,245 feet from the origin of the project's baseline.

The use of stationing in conjunction with a baseline allows great flexibility in writing descriptions of linear tracts that are of different widths in different locations along their total length. After describing the baseline, the writer can state the different widths to the left and right of that line between two stations (such as 25 feet to the left and 35 feet to the right between station 5 + 63 and station 6 + 18), or perhaps as one full width between stations 0 + 00 and 2 + 50 and a different full width between stations 2 + 50 and 4 + 75.

The use of stationing with offsets from the baseline allows a simple presentation of locative information defining terrain features or property corners. The example in Figure 2.5 documents the location of the corners of the building nearest the road by referencing how far along the baseline it is to a point that is radial or at right angles to the building corner. In the example, the baseline is a straight line for 234.36 feet where it begins to curve (identified by the letters *PC*, for *point of curvature*). The curve is said to be curving "to the right" (based on the line of travel when entering that curve) and has a radius of 239.13 feet. The curved portion of the baseline ends at 491.24 feet from the beginning or origin of the baseline, and this end point of the curve is labeled *PT* for *point of tangency*.

From the drawing, we know that a line drawn from the building corner labeled "B" to the radius center of the curved baseline will cross that base line at a point 454.35 feet, as measured along the baseline (both its straight and curved segments), from the baseline's beginning point. Point "B" is shown to be 69.68 feet from the baseline as measured along the radial line. In other words, building corner "B" is offset 69.68 feet to the left of Station 4 + 54.35.

A line from point "A" perpendicular to the straight baseline will intersect that baseline at a point 497.32 feet from the baseline's origin. Point "A" is 106.64 feet left of the baseline. In route surveying lexicon Point "A" is identified as being "at Station 4 + 97.32, left 106.64 feet." This approach is often referred to as a *station and offset* description.

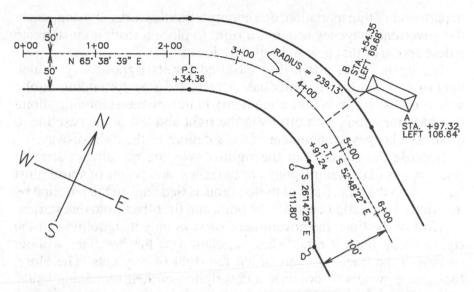

FIGURE 2.5 The stationing "offsets" are distances to a feature as measured perpendicularly to the baseline. In the case of curved baselines, the measurement is radial to the curve.

Right-of-way lines that are parallel to the baseline will have the same bearing as the baseline. In our example, line segment "C" to "D" can be described as beginning at Station 5 + 00, right 50.00 feet and ending at Station 6 + 00, right 100.00 feet. The distance between the end points of line segment "C" to "D" as measured along the baseline is 100.00 feet. The actual length of that line is 111.80 feet because of the varying distances of the end points from the baseline. The examples discussed here are intended to be introductory only. A thorough treatise on the stationing system would be a book in itself.

As with centerlines, the termination of a baseline at the boundaries of other properties must be checked to assure that the intended lands are included within the linear tract if the intersection is neither perpendicular nor radial.

2.4.5.2 *Government* Federal, state and local governmental bodies and certain quasi-public service providers can exercise limited rights to use or cross publicly or privately owned land. Often, the stated public purpose is linked to a specific system of describing the pertinent tracts.

2.4.5.2.1 *Rights-of-Way* Often, when a government entity refers to a right-of-way, it is describing a roadway, whether acquired in fee or as an easement. Whether the Federal Highway Administration, a state's department of transportation, or a county's public works division, when the government agency acquires a right to place a road, it must create a description of that new right-of-way location.

The agency will establish a baseline for its right-of-way, which may or may not be the centerline. A recorded map (providing public notice of the location and alignment) indicates the stationing along that baseline, along with offsets to the right and left of the baseline to establish changes in alignment of the sidelines of the right-of-way.

Because the sidelines of the right-of-way are not always straight lines, points of curvature, points of tangency, and points of spiral must be noted on the plan. Each of these points is tied back to the baseline by notation of the station opposite the point and the offset from that station.

Most of the time, the government surveys only its baseline and the right-of-way lines it establishes in relation to the baseline, without surveying the tracts through which the right-of-way cuts. Therefore, the private owners do not have a description of their remaining lands, only their original deeds less the area taken by the new right-of-way. The "remainders" of lots affected by such right-of-way acquisitions are approximate due to the fact that neither the frontages nor the sidelines of their properties emanating from that frontage have been surveyed in the process of creating or expanding the right-of-way.

As a result, the written description of the right-of-way really reflects only the acquiring agency's graphic depiction of the right-of-way, with dimensions and area calculated from deeds that may or may not be based on surveys. Therefore, when describing such subtractions from private lands, reference to the title and date of the recorded plan is immensely important in establishing intent. There may also be monumentation set as part of the right-of-way construction project, and, if recovered, these should also be mentioned in the written description. However, such monumentation should be checked carefully in the field, as it is not always set by standard surveying procedures, and in some states has been set by augurs mounted on backhoes, consequently only approximating the points meant to be marked.

2.4.5.2.2 Easements Government entities may also employ descriptions in relation to a baseline or centerline when acquiring easements along rights-of-way. If, for instance, a drainage easement or bridge maintenance easement or slope easement runs along the sideline of the right-of-way, the description of such an easement can be by reference to the offset from stationing on the baseline or centerline.

In Figure 2.6, we have a baseline marked in 25-foot increments from a starting point to the left of this site. The 10-foot-wide drainage

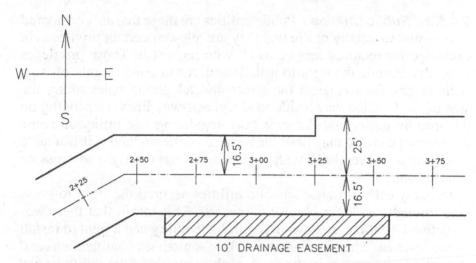

FIGURE 2.6 **The stationing along the baseline or reference line tells us how far we are from the start of the project, and offsets to the right or left from that baseline provide the framework for the geometric relationship between various features related to that baseline.**

easement along the side of the main right-of-way can be described several ways in relation to this baseline:

A drainage easement beginning at a point offset 16.5 feet south from Station 2 + 52 and running to a point offset 16.5 feet south from Station 3 + 60, being ten feet wide and south of this described line.

A 10' wide drainage easement running from Station 2 + 52 to Station 3 + 60, with its northerly line being offset 16.5 feet south of and its southerly line offset 26.5 feet south of the said baseline.

A drainage easement running parallel with and extending ten feet in width from the south line of the road right-of-way line between Stations 2 + 52 and 3 + 60, said right-of-way line being offset 16.5' to the right of said baseline.

Different agencies and description writers have developed innumerable variations on this theme over the years, some clearer and more understandable than others. Of course, the source of this baseline must be referenced somewhere in the description. Who established it and when? Has its location been recorded in a filed plat? Does the right-of-way have a name (perhaps it is State Highway 47 or County Route 313)? Is the origin (0 + 00) monumented or referenced to a physical, recoverable feature? The reader (or the reader's surveyor) must be able to recreate the line on the ground, and providing this kind of evidence is key to that effort.

2.4.5.3 Public Utilities Public utilities are those that are considered so essential to quality of life that they are allowed certain privileges in exchange for required service to all who request it. Those privileges generally include the right to install facilities in public rights-of-way, such as gas lines beneath the street and telephone poles along the side of the traveled way inside road right-of-way lines. Depending on the specific utility and the state laws regulating that utility, in some instances the utility may have the right to condemn land if it is unable to negotiate successfully with a landowner for the right to cross or acquire land.

In many early instances, public utilities secured the right to cross land through deeds for *blanket easements*. This meant that there was no defined location set down in writing; the utility had a right to install its facilities anywhere on the tract. While sometimes qualifications and restrictions did appear in the deed, such as requiring the utility to pay for any damaged crops or having to install fences to keep livestock from disturbing facilities, for the most part these were boilerplate descriptions, only occasionally referring to an attached sketch identifying the tract being crossed by the utility.

Once the utility facility is installed, however, a blanket easement no longer vaguely floats "somewhere" on the land. Many courts throughout the country have stated that the first installation of a facility establishes the definite location of the utility's right-of-way or easement. The width is less exact, but is generally set at an extent allowing maintenance and repair of the facility—again setting the size of the original facility as the reference. If an electric company installed a line of poles across a farm in 1939, each pole with a cross-arm carrying a total of four wires, the electric company cannot unilaterally later decide to install high-tension wire towers in the same easement. That would be an additional burden on the landowner, requiring a bigger easement than the one established by the original pole line and therefore requiring new negotiation with the landowner.

If, however, the deed allowed the electric company to install "such facilities as convenient and necessary to its operations," additional wires may be allowed on similarly sized poles, but wider cross-arms might not be permitted. The situation is resolved case by case, dependent upon facts at the time of the original conveyance, including contemporaneous standards of both language and utility construction methods.

Returning to the pole line blanket easement rather than the tower scenario, it is common for title companies to merely recycle old descriptions rather than update them to reflect current conditions. As a result, we find title searches with blanket easements rather than more specific descriptions of where the burdens lie on a particular property. This is where the surveyor has an opportunity to serve the public, both in providing a specific description to serve the client and in clearing out extraneous language of old descriptions. If no construction ever occurred, the blanket easement still is the only valid description. However, when appropriate, the surveyor can employ the linear tract form of description to identify the location of the easement over the property. A pole line or a series of sewer manholes can be used as the baseline or centerline. New lots divided from the original tract in locations that never experienced any construction based on the original blanket easement are no longer subject to that easement, due to the easement being fixed elsewhere upon construction of the facility. Verifying the actual physical location of a formerly blanket easement helps clear irrelevant references from the records of such sites.

2.4.5.4   *Private Easements*   There are times when one landowner must cross another person's land to access public roads. In such instances, there may be a private right to cross the neighbor's land that has never been formalized in writing. However, it is up to the writer of a description to include all such rights in the document, identifying all the factors to which a property is subject or with which it may be enjoyed. And so we should be describing the private roads. While it is possible that an agreement for the private easement has already been formalized in writing, lack of that definitive document does not mean we cannot describe private easements.

The physical centerline of the traveled way as located in the field is a suitable reference line, with a width that is agreeable to both the user of the private road and the owner of the land crossed by that private access way. The width established should be adequate to maintain and repair the private way. Merely reflecting the existing variable traveled width as the road bends and twists its way through a property may not be the most suitable means of establishing a private access easement.

2.5   COMBINED RECORD SYSTEM DESCRIPTIONS

The metes and bounds land record system formed the foundation for real property transfers and laws for generations, and therefore laws governing real property transfers evolved in a metes and bounds environment. Other record systems developed here in the United States to simplify land accounting, such as the USPLS and platted subdivisions, but these systems may also employ metes and bounds to describe specific tracts.

Such supplemental metes and bounds descriptions sometimes result from the desire to improve on the form of "legal description" that is established by the system used to create the parcel. A "legal" description is:

> A description recognized by law which definitely locates property by reference to *government surveys, coordinate systems or recorded maps;* a description which is sufficient to locate the property without oral testimony.[3] *[Emphasis added]*

In accordance with this definition, an acceptable legal description of Lot 34, Square B, Sunshine Estates, Hill County, Montana, is simply: "Lot 34, Square B, Sunshine Estates, Hill County, Montana," although it is strongly recommended to include additional information guiding the reader to where the plat for Sunshine Estates may be found in the public records. The abbreviated description can be supplemented with a metes and bounds description of encumbrances such as later easements or additional road right-of-way widening not appearing on the referenced platted subdivision.

Turning to the USPLS, the full and accepted legal description of the southeast quarter of Section 5 within the Township in the third tier north of the baseline and the fourth range west of the Boise Meridian in Ada County Idaho is: "The SE1/4 of Section 5, Township 3 North, Range 4 West, Boise Meridian, Ada County Idaho."

Survey plat standards and specifications that require the "legal description" to appear on the face of the plat *do not* necessarily require that a metes and bounds description be placed on the plat if an abbreviated legal description, as shown in these examples, suffices to allow a surveyor to locate the tract on the ground.

Metes and bounds descriptions of parcels created under one of the other land record systems can be developed, but only as a supplement to the legal description and not to alter the plain meaning of the USPLS or platted subdivision reference. If brevity requires omitting important locative and identification information, then such a supplemental description should not appear on the face of a survey plat. In such

---

[3]*Definitions of Surveying and Associated Terms*, American Congress on Surveying and Mapping, revised 2005.

instances when space is at a premium, the description inevitably is stripped of its usefulness in reporting evidence and becomes a simple recitation of the bearings and distances that are already present on the plat.

Whatever means of land description is employed in any of the land record systems, the ultimate objective is clear, concise, and accurate written documentation of the physical and record location of a parcel. A properly written land description should provide sufficient evidence that any competent land surveyor could recover the boundaries of that parcel even if the original survey plat or subdivision plat has been lost.

CHAPTER 3

DIRECTIONS

Descriptions of land take many forms, and in this chapter we address some of the forms that include references to directions along the lines of a boundary. Mathematical and geometrical principles provide the backbone of these approaches to describing the shape of a tract of land; while we will introduce those principles here, a general surveying or geometry text will provide details beyond our discussion of the form and application to land descriptions.

3.1 ANGLES

When describing the shape, size and relative position of tracts by straight and curved lines, angular measurements describe the relationship of straight lines. There are two distinct types of angles associated with real property boundaries. A plat or deed may present a measure of the angular relationship between two intersecting boundary lines. This directly states the orientation of the lines to each other and is silent concerning any orientation to the Earth. Alternatively, as we will see in section 3.2, the plat or deed may present a measure of the relationship of a boundary line to a reference line or meridian, such as "north." This describes a boundary's relationship to that reference line, but requires some mathematical exercise to determine one boundary line's relationship to another.

3.1.1 General

Angles are a measure of the relationship between two intersecting lines. Because two intersecting lines define a plane, it follows that an angle can exist in only one plane. Land surveys take place on the surface of a round planet. The multiple angles that are measured and reported during the course of a real property boundary survey must be coplanar, meaning within the same reference plane. The horizontal plane adopted for real property surveys and descriptions is a function of the projection system used to map the work.[1] Because survey plats are flat surfaces, whether printed on paper or displayed on a computer screen, we will consider only horizontal coplanar lines in this limited discussion of angles.

3.1.2 Interior Angles

An angle is said to be *interior* when it is on the inside of a closed figure, enclosed by the exterior lines of the polygon. All of the interior angles are labeled "A" on the sketch in Figure 3.1.

Interior angles are useful both during the initial traverse and in analysis of the consistency of the data collected. Control traverses, as opposed to real property boundaries, rely on the basic geometrical fact that the sum of all the interior angles of a polygon will always

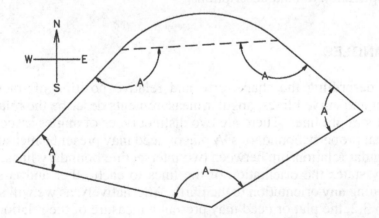

FIGURE 3.1 For the purpose of computations, the angles measured at curved boundaries are to the chord of the curve.

---

[1] See Chapter 4.

equal *(180°)* times *(the number of sides in the polygon minus two)*, as expressed in the well-known formula below.

$$\sum A_n = 180 * (n - 2)$$

"A" is the sum of all the interior angles in degrees, and "n" equals the number of sides of the closed polygon. The closed polygon in Figure 3.1 has five (5) interior angles. Therefore, the sum of all the interior angles in this case is 540 degrees, being the result of multiplying 180 by (5 minus 2).

In actual practice, the sum of the angles *measured* during a traverse will never equal the theoretical value dictated by the equation because of measurement errors. The difference between the theoretical sum and the measured sum of the angles is the "failure to close." This sum is often incorrectly identified as the "closure error." The "failure to close" is not the angular measurement error of the traverse; it is the arithmetic sum of all the measurement errors of each angle measured. Some of the *measured* values may have been greater than actual angle and some may have been less, thereby offsetting each other.

For more involved mathematical discussion of the process of surveying, the reader is invited to read elementary surveying texts for information that will not be covered here.

As an example of the use of interior angles in descriptions, it is not uncommon to find property lines defined as running at right angles to each other, after defining a base or reference line from which the description begins. The following is an actual description as an example of this approach.

> BEGINNING at a point in the north line of Main Street, 200 feet west of the point of intersection of the north line of Main Street with the west line of Front Street, and running thence:
>
> Along the north line of Main Street, 25 feet, thence:
>
> At right angles to the said line of Main Street, 150 feet, thence;
>
> Along a line forming an interior angle of 100 degrees from the prior course, 35 and a half feet, more or less, thence
>
> Along a line parallel with the second course, 175 feet, more or less, to the point and place of beginning.

While this description may not mathematically close perfectly, the writer's intent regarding the shape of this tract is clear from the language of the document.

3.1.3 Exterior Angles

An angle is said to be *exterior* to a closed figure when it lies outside the closed polygon. Because the space encircling any given point contains a full 360 degrees, a given exterior angle of a polygon is equal to 360° minus the interior angle at that particular point. Stated another way, the relationship between the two lines forming that exterior angle is defined by an arc located in the space outside of the closed polygon.

Figure 3.2 includes a chord line defining the limits of the curved section of the tract's perimeter, with the exterior angles running from the outside of the tract to the outer edge of that chord line. Such an approach was often used when the drafter of the description either did not have confidence in himself or in those following his description regarding the accuracy and precision in defining the curved boundary of a tract. This approach also permits a rough check of the geometry of the entire perimeter, by providing a closed polygon with established angles and line segment lengths that must conform to basic geometric principles in order to close properly.

3.1.4 Deflection Angles

An angle is said to be a *deflection angle* if it is documented by reference to an extension of the "line of travel," or the consistent direction in which the survey and description progress. If the next line in the description is to the right of the line of travel, the angle is reported as

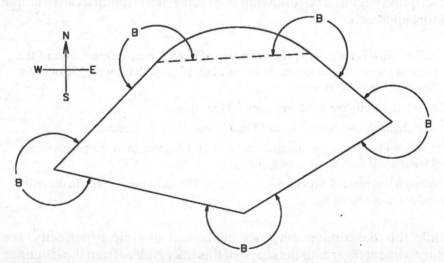

FIGURE 3.2 Angles at curved boundaries are shown to the chord of the boundary.

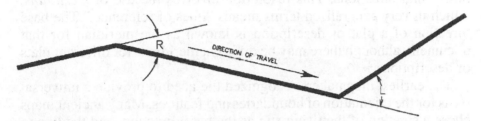

FIGURE 3.3 Deflection angles are described as if the reader were moving along the line and turned to the new line by "deflecting" the line of travel.

a *right deflection,* while *left deflections* refer to a leftward change in direction of the line of travel. In the example shown in Figure 3.3, "R" signifies a right deflection and "L" a deflection to the left. If deflection angles are used in describing a closed polygon, then geometrically the sum of the deflection angles for that polygon will be 360 degrees, with right deflections considered to be positive values and left deflections considered negative.

3.2 MERIDIANS

Angles define the relationship between lines, but do not orient those lines to the larger environment. Meridians provide a larger framework to tie numerous tracts and parcels to each other in the same reference system. The description writer should always define the reference system or meridian, which can be established in an infinite number of ways. An Earth-based (geodetic) foundation is the most effective and recommended mechanism for defining a meridian.

3.2.1 General

The use of internal, external, and deflection angles in describing a real property parcel is limited, as they define configuration but not necessarily the location or orientation of the tract in relation to any other object on the face of the Earth. Boundaries documented by survey plats and descriptions that rely on angles alone are difficult to recover. Proper retracement of such boundaries requires the recovery of at least two of the corners reported on the original survey plat or in the description as a means of orienting the search for the remaining corners. For anyone other than individuals already familiar with the parcel to be able to orient a tract in the way intended, we need a reference system of known places and directions. This resolution involves the use of *meridians,* which in very generalized terms means "lines of reference." The base direction of a plat or description is known as the meridian for that document, although there may be different meridians for different plats or descriptions.[2]

The earliest mapmakers recognized the need to provide a universal basis for the orientation of boundaries and features. Many ancient maps chose a sighting of the rising sun as the base direction, and the line to the sun became the reference meridian to which all other locations on the map were referenced.

Advances in astronomy, however, provided a more precise base direction than the sun. Observers of the night sky quickly detected an apparent center of rotation of celestial objects that did not require seasonal adjustments to arrive at a base direction, as was the case with relying on the sun. The direction "north" became the universal base direction or meridian for maps and, consequently, for real property boundary descriptions based on those maps and surveys.

"North" can have many different definitions, and it is important to understand the original intended definition before attempting to interpret a written description of property that includes references to cardinal directions within its courses. The attempt to describe three-dimensional boundaries on a round planet using terms and figures derived from flat maps has resulted in a variety of possible north-defined meridians, which we briefly define in the following sections.

3.2.2 True North

True north is a three-dimensional term.[3] The axis of the Earth's rotation is a line segment that ends at the planet's poles. Any geometric plane that contains this axis of rotation is defined as being oriented to true north. A true north geometric plane will intersect the surface of the Earth along a line that is also known as true north (Figure 3.4). Unfortunately,

---

[2]See Chapter 4.

[3]Stephen V. Estopinal, *A Guide to Understanding Land Surveys*, 3rd ed. Hoboken, NJ: John Wiley & Sons, 2009.

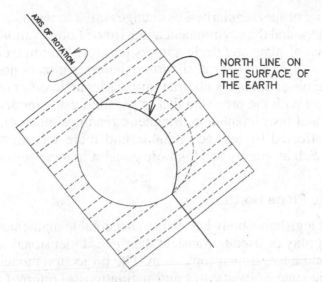

FIGURE 3.4 Any plane that contains the axis of rotation of the Earth will intersect the surface along a true north line.

the Earth wobbles slightly on its axis, so true north lines are not entirely constant. These variations are slight and were, until the development of the very precise measurement systems we have today, undetectable.

3.2.3 Astronomic North

Astronomic north is a three-dimensional term. The determination of north based on observations of the apparent center of the earth's rotation using the stars had long been thought to identify true north as well. But because such observations must be made while standing on the surface of the Earth, gravitational anomalies can generate observational errors. These errors were not understood until the 20th century. Astronomic north, unless corrected for these small gravitational anomalies, will result in a meandering line that closely approximates true north.

3.2.4 Magnetic North

Magnetic north is a three-dimensional term. It is defined as the determination of north based on the magnetic field of the Earth. The difference between the magnetic lines of force and true north is called the *magnetic declination,* which is different for every point along a line (point specific) and is not constant. The poles of the magnetic lines are moving, more or less regularly. The changes are a function of the date, the time of day, and the environment at the time of observation. Correction for much of magnetic declination is possible but imprecise. *Even when corrected for magnetic declination*, determination of north based on magnetic observations is not true north. Magnetic observations are not consistent with the precision that is possible using modern survey procedures and instrumentation. Magnetic readings on a compass are also often affected by less accountable and more inconsistent local conditions such as gravity and high ore content in local soils.

3.2.5 State Plane North

Each state's legislative body has defined acceptable mathematical procedures that may be used to translate the three-dimensional surface of the Earth onto a two-dimensional map. The projection model adopted by a specific state in its statutes and regulations is referred to as the *state plane projection system* or *state plane coordinate system* (SPCS). Directions in the state plane system are oriented to true north.

State plane north lines are parallel lines, and two-dimensional Cartesian coordinates are used to define locations. See Chapter 4 for additional discussion about the various means of describing the relationship between straight lines on a curved earth, also call *projections*.

3.2.6 Assumed North

Assumed north can be a two- or three-dimensional term. The most common use of assumed north is two-dimensional and associated with a tangent plane projection.[4] The original surveyor may simply define the direction of a line between two points in the survey as having a particular bearing and determine directions based on that assumption. Subsequent surveyors may reference their work to the same reported direction between two recovered points from the previous survey. In such a case, the later surveyors are assuming and following the meridian of their predecessor.

As in any reference system, no problems arise as long as everyone is using the same system. As soon as someone decides to change the assumption regarding in which direction north lies, some kind of transformation or conversion is necessary to keep all the parts of the system in the same relationship to each other. We face similar situations when work entirely within the metric system or entirely within the

---

[4]See Chapter 4.

English measurement system must be converted into the other system's units. There is no loss of accuracy as long as a means of accomplishing the transformation or conversion is available. Therefore, identifying where "north" is, whether in relationship to the stars or to a prior deed, provides a benchmark for future readers of the description.

3.3 BEARINGS

Each line in a survey can be described by reference to other lines in the survey (i.e., interior and/or exterior angles) or by reference to a meridian. North, in all its variations, has been universally adopted as the meridian of choice in describing boundaries (although geodetic and other technical surveys may reference directions from south). The most common method of describing the angular relationship of a line with the meridian is the *system of bearings*. For the purposes of this discussion, we will consider only the two-dimensional relationship of mapped lines.

3.3.1 North-South Reference Lines

Bearings can describe the angle formed between a line and an identified meridian in terms of a quantity of degrees east or west of that meridian. We describe a line as if the writer/reader were standing on the point of intersection of the meridian and the line being described, facing along the meridian and then turning toward the east or west to then face along the line of travel.

In Figure 3.5, an observer facing to the north while standing at the intersection (Point C) of the described line BA with the meridian would have to turn toward the east (turn to the right) 47 degrees 56 minutes 58 seconds in order to then face point "A" along the described line. If that same observer standing at Point C were to start by facing to the south, he or she would have to turn toward the west (turn to the right) 47 degrees 56 minutes 58 seconds to face point "B" along the line (which can now be referred to as line AB due to the new direction of travel). North-east and south-west bearings are considered as positive or right turns when traveling clockwise around a described polygon, while north-west and south-east bearings are left turns.

Notice that if the numeric values of a bearing are the same, north-east and south-west bearings describe the same line. The only change is the direction of travel, which is only pertinent if a person were actually moving along the line. An observer moving toward point "A" from

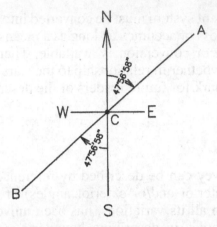

FIGURE 3.5 The north-south reference line (shown with an arrow head for orientation) represents a meridian. The ends of the line to be described are marked as "B" and "A", and the line intersects the meridian at Point C.

point "B" along line AB is traveling northeast. Moving toward point "B" along line AB means that the observer is traveling toward the southwest.

Consider the case demonstrated by the sketch in Figure 3.6.

An observer standing at the intersection (Point F) of the line being described (CD) with the meridian and facing to the north would turn toward the west (to the left) 57 degrees 46 minutes 03 seconds in order to face point "C." If standing at that same intersection but starting by

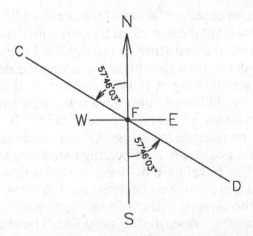

FIGURE 3.6 The meridian is the same as in Figure 3.5, but a different line is being described.

facing to the south, that same observer would have to turn toward the east (to the left) 57 degrees 46 minutes 03 seconds in order to face point "D."

All bearings must lie within a particular quadrant (or quarter) of a full circle (360 degrees). Lines running exactly in one of the four cardinal directions (north, south, east, west) can be expressed in three ways, depending on the writer's preference. For example, a line running due north (with no deflection either east or west of exact north) can be described as "North" or "North 0 degrees East" or "North 0 degrees West." A line running due (or exactly) west can be expressed as "West" or "North 90 degrees West" or "South 90 degrees West." The direction for a line running in a cardinal direction is most commonly reported as simply "North," "South," "East," or "West." Bearings are always between the values of zero and 90 degrees.

3.3.2 East-West Reference Lines

The use of the east-west meridian is rare in modern land surveys, and was only slightly more common in precolonial and colonial descriptions. If a line running east-west is defined as the reference meridian, then lines can be described as being north or south of that meridian. Figure 3.7 indicates a line that could be described as running East 56 degrees 15 minutes North in such a system.

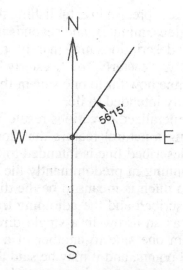

FIGURE 3.7   Here the reference meridian runs east and west rather than north and south.

3.3.3 Reversing Directions

One of the benefits of using bearings in describing lines is that reversing directions along a line does not involve any mathematical computations, due to the same intersection relationship between the line and the reference meridian in both forward and reverse directions. This is in accordance with the basic geometric principles that (1) a straight line has an angular value of 180 degrees and (2) any other line intersecting that line creates equal angles on opposite sides of the point of intersection. Therefore, the reverse direction of a north-east line is south-west and vice versa. The reverse direction of a north-west line is south-east and vice versa.

This reverse relationship allows a direction in the descriptive narrative that was not anticipated by the author of a survey plat on which the original description was based. In some instances, reading and running a description in the reverse direction from which it was written allows a later surveyor to retrace and recover the intended lines of a boundary, particularly when some of the courses in a description are ambiguous, generalized, or even missing or possibly in error. The courts in this country commonly accept the practice of reversing directions in a description to allow the removal of ambiguities, but never to introduce a new meaning to accepted meaning and intent.

3.3.4 Generalized Directions

Some descriptions are less precise in establishing the direction of lines, and do not include degrees, minutes, and seconds in their statements of bearings. Instead, such descriptions may merely state that a line runs in a "northerly," "southerly," "easterly," or "westerly" direction, leaving it to the reader to determine how far to one side or the other of a cardinal direction the line is truly intended to lie.

Sometimes these generalized directions result from the fact that the course is running along a natural feature that does not run perfectly straight; instead, the described line is intended to follow the meanderings of that feature running in predominantly the same general direction. For example, if a ditch is meant to be the dividing line between the property being described and the adjoining tract, that ditch is not likely to run straight as an arrow in a single direction. However, its slight wanderings from one side to another of a single direction are defined by its two end points, and it may be said to run "northwardly" or "southwestwardly" or "generally in an easterly direction."

When describing a natural feature that is intended to be the boundary, additionally defining a tie line between its origin and terminus with a definite bearing and distance can provide some certainty, helping future surveyors recover lines when natural features have been disturbed or destroyed. The direction and/or distance, however, will never overcome the intent of that natural feature to serve as the boundary.

> ...thence approximately 1210 feet along the centerline of Vine Creek to a point in the southerly line of lands now or formerly of Edward Rush, with a tie between the origin and terminus of this course bearing North 56 degrees 27 minutes 12 seconds West, 987.32 feet...

3.4   CURVED LINES

Not every line in a boundary is straight, or even generally straight, and often curves define the shape of a tract. Most of these are simple circular curves (curved lines with a single center point and a fixed or constant radius). However, in some instances, primarily in relation to railroad or highway rights-of-way, we do encounter spiral curves (curves with a constantly changing radius).

The mathematical computation of curves will not be covered here. However, enough elements of a curve must be included in a description for subsequent surveyors to follow the intent of the curved line's location, length, shape, and direction. Although it is possible to recompute and thus recreate a curve from only two elements,[5] citing at least three elements provides the soundest basis for reestablishing a curve, although more may be included.

Among the possible descriptive elements of a curve are:

- The direction of the curve along the line of travel (curving to the right or to the left of the reader's progress along the tract's described perimeter)
- The radius (usually depicted on maps and surveys as R)
- The arc length, or distance traveled along the arc (often abbreviated on maps and surveys either as A for "arc" or L for "length of curve")
- The interior or central angle of the curved line (between the radial line from the beginning of the curve to the radius point and the radial line from the end of the curve to the radius point, abbreviated

---

[5]This is true only in the case of tangent curve relationships.

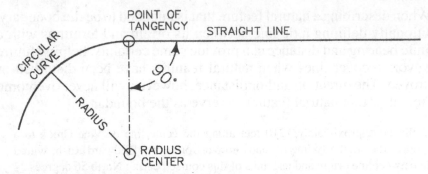

FIGURE 3.8   When the angular relationship between the radius center and the straight line at the beginning or end of the curve is a 90 degree intersection, it is termed "tangent". Any other angular relationship is nontangent.

on maps and surveys either as D for "delta" or change in the angle or its Greek equivalent, δ).

• The bearing and distance of the chord connecting the two ends of the curve

Additional descriptive elements clarify the relationship between straight and curved lines of a tract, and can prevent future misinterpretation of the boundary's configuration. There are a number of questions we should anticipate when describing curved lines, with some examples following.

Is the curve tangent to a line at either of the curve's ends (by "tangent" we mean that the angle between that straight line and a line to the radius point of the curve is 90 degrees, as shown in Figure 3.8)?

Is either the starting or ending point of the curve a point of compound curvature; does another curve connect at the ending point of the first, running in the same direction as the first curve?

Is either the starting or ending point a point of reverse curvature; does the next curve change direction, turning either right or left, in relation to the first curve? Figure 3.9 illustrates a variety of curve relationships.

Note the distinct geometric relationships between the curved courses and the courses preceding and following them in the following two partial descriptions.

Description 1

(3) Along the northerly right-of-way line of County Route 518, South 58° 24′ 13″ East, 69.83 feet to a concrete monument found at a point of curvature; thence

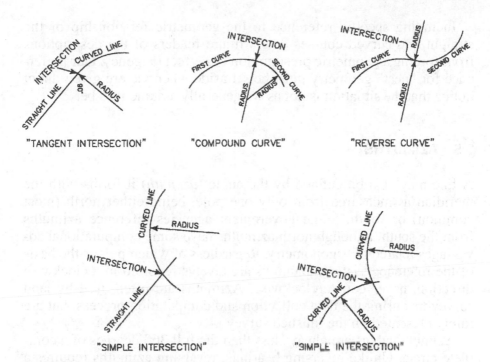

FIGURE 3.9 Curved lines have many relationships between each other and straight lines. These are the most common examples.

(4) Along curve to the right connecting the northerly right-of-way line of County Route 518 with the westerly right-of-way line of Irondale Road (33 feet wide), said curve having a radius of 50.00 feet and an arc distance of 78.54 feet, to a capped iron pin set at a point of tangency (this curved course having a central angle of 90° 00′ 00″, and a chord of South 11° 24′ 13″ East, 70.71 feet); thence

(5) Along the said westerly right-of-way line of Irondale Road, South 33° 35′ 47″ West, 61.08 feet to a capped iron pin set at an angle point; thence

Description 2

(14) Along the common line between Lots 4 and 5, South 75° 16′ 10″ East, a distance of 137.28 feet to a concrete monument found at a non-tangent point of curvature; thence

(15) Continuing along the common line between Lots 4 and 5, and along a curve to the right having a radius of 1157.00 feet, an arc distance of 222.38 feet (this curved course having a central angle of 10° 55′ 05″ and a chord of South 63° 11′ 30″ East, 222.04 feet) to a ³/₄-inch iron pipe found at a nontangent angle point; thence

(16) Continuing along the common line between Lots 4 and 5, South 35° 01′ 44″ East, 12.39 feet to an angle point; thence

Including specific reference to the geometric relationship of the straight and curved courses deters future readers of the descriptions from forcing a geometric presumption of perfect tangency. Our preference for "neat" geometry must be set aside when we are given direct notice that the situation is not as we generally assume it to be.

3.5 AZIMUTHS

A line may also be defined by the angle (*azimuth*) it forms with the meridian as measured from only one pole, being either north (most common) or south. Some government agencies reference azimuths from the south, although north azimuths have some computational advantages related to trigonometry. Regardless of which pole is the basis of the reference system, azimuths are always reported in a clockwise direction, never counterclockwise. Azimuths are often used by land surveyors during the data collection and computation process, but are rarely presented on the finished survey plat.

Azimuth values are always less than the full 360 degrees of a complete circle. Unlike reversing bearings, reversing azimuths requires a mathematical computation involving the addition or subtraction of 180 degrees from the original azimuth value to derive the reversed azimuth value. When the original azimuth value is less than 180 degrees, then 180 degrees are added to that original value to derive the new reversed azimuth value. When the original azimuth value is greater than 180 degrees, then 180 degrees are subtracted from the original azimuth value to derive the reversed value.

The heavy dashed arrow in Figure 3.10 represents the meridian from which the azimuths in this north-based system will be based. This sketch shows the values, in terms of north azimuths, of the example lines for bearings. The angles are all "reckoned" as to the right, clockwise, from the meridian.

The direction of line BA can be described as an azimuth of 47 degrees 56 minutes 58 seconds from north. If we were to turn ourselves around and walk back on that line to our starting point, the direction of line AB would have an azimuth of 227 degrees 56 minutes 58 seconds from north (being the sum of 180 degrees plus 47 degrees 56 minutes 58 seconds, since we are now headed in a southwesterly directly).

If we were to reverse the reference meridian and use a south azimuth as our line of reckoning, line BA would have an azimuth of 227 degrees 56 minutes 58 seconds from south (180 degrees plus 47 degrees

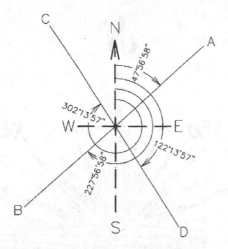

FIGURE 3.10   All angular measurements or azimuths in this diagram are described clockwise (to the right) from the north-south line representing the meridian.

56 minutes 58 seconds), and line AB would have an azimuth of 47 degrees 56 minutes 58 seconds from south. This example also illustrates the geometric principle of opposite angles being equal.

3.6   COMPASS DIRECTIONS AND HEADINGS

The use of bearings developed along with improved nautical navigation, and remnants of this traditional seafaring method of describing directions have persisted on land in some areas. Directions on a compass originally were related to the directions of the 8 major winds, the 8 half winds, and the 16 quarter winds that affected sailing ships on the high seas, creating 32 defined directions. Thus, the directions shown on a 32-point compass were often called *headings,* referring to the direction that the ship was pointed for its intended line of travel. When heading in a cardinal direction (north, south, east, or west), the term *due* (such as "due north") indicates strict adherence to that direction, with no deflection to the right or left of it.

The 32-point compass (as in Figure 3.11) divides the 360 degrees of a full circle into $11\frac{1}{4}$ degree (11° 15′) increments. All directions are related to "north," which has an assigned value of "zero degrees," thereby providing a north-based meridian. The degrees designated in the following table are north-based azimuths.

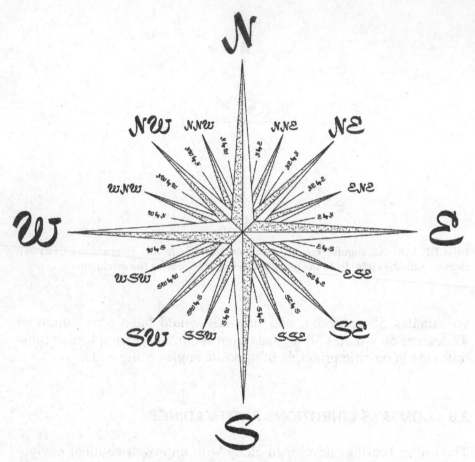

FIGURE 3.11 Compass Rose. A mariner's compass rose was painted on the face of card attached to a magnetic bar floating or suspended in a box aligned with the keel. As the vessel turned, the card would move relative to the box and the helmsman could read the line of travel of the ship.

Direction	Abbreviation	Degrees (degrees, minutes)	Decimal Degrees
North	N	0°	0.00°
North by East	NxE	$11^{1}/_{4}°$ (11° 15′)	11.25°
North Northeast	NNE	$22^{1}/_{2}°$ (22° 30′)	22.50°
Northeast by North	NExN	$33^{3}/_{4}°$ (33° 45′)	33.75°
Northeast	NE	45°	45.00°
Northeast by East	NExE	$56^{1}/_{4}°$ (56° 15′)	56.25°
East Northeast	ENE	$67^{1}/_{2}°$ (67° 30′)	67.50°
East by North	ExN	$78^{3}/_{4}°$ (78° 45′)	78.75°
East	E	90°	90.00°

Direction	Abbreviation	Degrees (degrees, minutes)	Decimal Degrees
East by South	ExS	$101\frac{1}{4}°$ (101° 15′)	101.25°
East Southeast	ESE	$112\frac{1}{2}°$ (112° 30′)	112.50°
Southeast by East	SExE	$123\frac{3}{4}°$ (123° 45′)	123.75°
Southeast	SE	135°	135.00°
Southeast by South	SExS	$146\frac{1}{4}°$ (146° 15′)	146.25°
South Southeast	SSE	$157\frac{1}{2}°$ (157° 30′)	157.50°
South by East	SxE	$168\frac{3}{4}°$ (165° 45′)	168.75°
South	S	180°	180.00°
South by West	SxW	$191\frac{1}{4}°$ (191° 15′)	191.25°
South Southwest	SSW	$202\frac{1}{2}°$ (202° 30′)	202.50°
Southwest by South	SWxS	$213\frac{3}{4}°$ (213° 45′)	213.75°
Southwest	SW	225°	225.00°
Southwest by West	SWxW	$236\frac{1}{4}°$ (236° 15′)	236.25°
West Southwest	WSW	$247\frac{1}{2}°$ (247° 30′)	247.50°
West by South	WxS	$258\frac{3}{4}°$ (258° 45′)	258.75°
West	W	270°	270.00°
West by North	WxN	$281\frac{1}{4}°$ (281° 15′)	281.25°
West Northwest	WNW	$292\frac{1}{2}°$ (292° 30′)	292.50°
Northwest by West	NWxW	$303\frac{3}{4}°$ (303° 45′)	303.75°
Northwest	NW	315°	315.00°
Northwest by North	NWxN	$326\frac{1}{4}°$ (326° 15′)	326.25°
North Northwest	NNW	$337\frac{1}{2}°$ (337° 30′)	337.50°
North by West	NxW	$348\frac{3}{4}°$ (348° 45′)	348.75°

CHAPTER 4

MAP PROJECTIONS

4.1 GENERAL

The history of maps and mapmaking predates writing. One can imagine a Neolithic huntsman drawing a sketch in the sand to communicate the location of water or game. As villages grew into city-states, the people began to assign farm plots, plan for defense, and to undertake a myriad of other activities associated with civilization. Maps began to catalog the reach of the known world.

Clay tablets, stone slabs, and paper on which people drew maps were flat, and so, everyone thought, was the world. The areas covered by these early maps were small, and the measurements used to dimension the features so crude that many centuries passed before two facts became evident:

1. The Earth is round.
2. Drawing a round surface to scale on a flat surface does not work for large areas.

It is not possible to draw a map of a round planet on a flat surface without distortion. Globes solve the problem of distortion for large areas, but the scale possible with a globe is small and the image lacks much detail (not to mention that carrying a globe around while traveling is highly inconvenient). The science of cartography overcame the distortion problem by developing methods of mapping known as *map projections,* or approaches to depicting ("projecting") the curved surface of the Earth on a flat surface. Although accomplishing it differently, all of the various projection models take into account two major differences between a round planet and a flat plane.

Examine a globe and you will notice that the true north lines (lines of equal *longitude,* also called the meridians) are straight but not parallel. All lines running north-south converge at the poles. All maps utilize a Cartesian coordinate system to plot features on the flat paper surface. But Cartesian coordinates define all north-south lines as both parallel and straight. The Cartesian coordinate system is a method of expressing the location of a point, on a flat plane, in terms of its position relative to two perpendicular lines commonly known as the X and Y axes. Any point may be described by the spot at which lines drawn from that point, those drawn lines being perpendicular to the horizontal (east-west) X and vertical (north-south) Y axes, intersect those two axes. In a two-dimensional system, Cartesian coordinates are a set of *x* and *y* values for any given point correlating to "east" and "north" values as offsets from a given point of origin (being the intersection of the X and Y axes) on a flat surface. These differ from geodetic coordinates that accommodate the curved surface of the Earth (which is actually more ellipsoid than spherical in shape). A three-dimensional Cartesian or geodetic system merely adds a value for *z,* the elevation above a defined plane or reference surface, a value with which we will not be concerned in this discussion. Figure 4.1 represents the basis for a Cartesian coordinate system in two dimensions, illustrating the relationship between the grid lines.

On a globe we can see that all east-west lines (lines of equal *latitude*) are parallel, but, except for the equator, these east-west lines are curved. In contrast, in a rectangular Cartesian coordinate system all east-west lines are parallel and straight. Projection systems, therefore, must distort distances between locations and convergence of the meridians in order to map an area of the round planet on a flat map surface.

Bearings and azimuths of straight lines are defined by the angle formed by the intersection of the line of observation with a meridian. As we follow lines of longitude from the equator toward either pole of the Earth, the distance between these lines becomes shorter, and eventually longitudinal lines converge at the poles. In the northern hemisphere, the result is that a tract of land with longitudinal lines as its east and west boundaries and with parallel lines as its north and south boundary lines will not be in rectangular form. Instead, the northern boundary will be shorter than the southern boundary line due

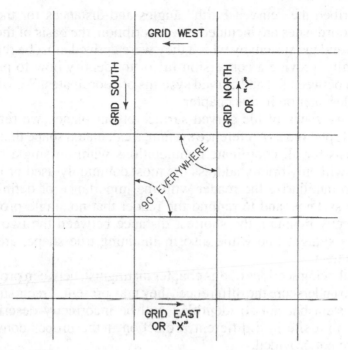

GRID WEST

GRID SOUTH

GRID NORTH OR "Y"

90° EVERYWHERE

GRID EAST OR "X"

FIGURE 4.1 The left edge of the mapping space was typically assumed as the "Y" axis and the bottom edge the "X." Even modern computer aided drafting programs use this type of control.

to the convergence of meridians. The convergence of the meridians means that straight lines, with the exception of the Equator and north-south lines, will not have a constant value for bearings or azimuths throughout their lengths. In contrast, Cartesian coordinates define all straight lines as having a constant bearing through their length. The numerical difference between true north at any point in a projection and north as presented in Cartesian coordinates is called the *mapping angle*.

Map projections are the processes by which we can transfer the coordinate positions from one surface to coordinate positions on another surface. Just as there are multiple means of defining "north" (true, magnetic, assumed, record, grid, geodetic, etc.), there are a numerous systems of assigning coordinate values to a specific point, each set of coordinates for any given point being unique within that given system so that no two points occupy the same space.

The difficulty of using a single series of coordinates to define a property boundary is that the coordinates allowing us to calculate "flat Earth" angles and distances do not correlate exactly to the coordinates that described the "curved Earth" angles and distances for the same lines. If coordinates are included in a description, the basis of those coordinates and the system by which they were derived must be reported. Additionally, provide a conversion factor to identify how to properly transition between flat and curved systems of coordinates. We will look at examples later on in this chapter.

When we speak of the curved surface of our planet, we refer to a computed spheroid as a reference system, a calculated shape that serves as the basis for all coordinate computations within a single system. This text will only briefly address the most commonly used projection models to familiarize the reader with the importance of defining the reference systems, and to remind the reader that no single projection will perfectly maintain the shortest distance between the two points defining a straight line while also maintaining true shape, area, and scale of a tract.

The following sections of this chapter distinguish between projection systems to underscore the differences they may present. Tracts surveyed in one system but not so identified will be incorrectly described in reference to a site in a different projection if the proper conversion factors are not provided.

4.2 PROJECTIONLESS MAPS

Maps can be made by simply drawing terrain features on paper without concern for the curvature of the Earth, provided that the area being mapped is small enough and the precision of the directions and distances is relatively crude. *Projectionless maps* deal with the convergence of the meridians and with the curved surface of the Earth by simply ignoring these facts. For this reason, projection-less maps do not provide a *scale factor*, or means of converting the distances shown on the map to actual distances measured on the datum that serves as the basis for the map. It is only when a scale factor is reported as "1.0" that map distances and ground distances are identical. We will refer to the situation of scale factor being equal to 1.0 as "unitary scale factor."

Maps drawn in conformance to one of the various projections generally provide a means of converting reported distances to lengths within the datum or reference system of the projection. Descriptions for a property depicted on a map based on a particular projection or state plane coordinate system should provide that scale factor as well as the name of the projection and a reference set of coordinates for at least one point on the boundary in order to pin the property down on the surface of the Earth.

4.2.1 Government Land Office (GLO) Plats

In 1812, the United States established the Government Land Office (GLO) to be responsible for surveying, platting and selling all public domain lands, or those tracts owned by the federal government and not yet sold or granted to private entities and individuals. The lands so administered by the GLO were tracked through a rectangular cadastral system created by the Land Ordinance of 1785, since modified by a series of *Letters of Instruction* and *Manuals of Instruction*, and known as the Public Land Survey System (PLSS).[1]

The PLSS is an idealized system, dividing the curved surface of our portion of the globe into a grid system based on townships and sections. The surveying process requires control lines, in the form of a *baseline* running east-west and a *principal meridian* running north-south. These lines intersect at what is known as an *initial point*. These elements correlate, respectively, to the X axis, Y axis, and point of origin as described earlier in this chapter. The full process of surveying PLSS lands is beyond the scope of this text.

However, the history of implementing the PLSS tells us much about the root cause of various problems encountered by later retracers and subdividers of original PLSS surveys. GLO Public Lands Surveys were typically accomplished by the surveyor observing a magnetic bearing, adjusting for declination, and measuring distances between corners with a chain. The direction of each line was independently observed. The observed bearings and the distances measured were then plotted on a map.

The resulting GLO Official Plat was a projectionless map. Nothing was done to account for distortions. Closures were not computed, but instead were field observed and adjusted. More often than not, closures were not even attempted. Each government surveyor was given a set of instructions on how to survey PLSS sections. How the work was actually done often deviated from those instructions.

4.3 CONFORMAL PLANE PROJECTION

Conformal plane projections are a category of mapping systems that are best suited for state-sized mapping projects. These projection systems are "conformal" because angles measured between lines on the Earth's surface are not distorted during the projection process. There is

---

[1] See section 2.2 of this text regarding the United States Public Land System (USPLS), the land record system for which the PLSS surveying methods were applied.

a large variation (mapping angle) between true north and grid north.[2] Conformal maps show shape accurately (they "conform" to the shape of land masses), but tend to distort at the top and bottom as the image moves away from the equator, with increasing inaccuracy in scale as the distance from the equator increases. Both the convergence of the meridians and the distances along the curved surface of the Earth are manipulated using a rigorous mathematical adjustment model. Such models are designed to cover a specific shape and extent of the Earth's surface area while minimizing the distortions involved. Tangent plane, Lambert, transverse Mercator, and state plane projections are all conformal plane projections.

4.3.1 Tangent Plane Projection

Instead of observing the direction of lines independently of one another, one might establish a direction or meridian as a reference and measure the angles formed at the corners of a tract in relation to that direction or meridian. Plane geometry could then be used to analyze the angles and distances between the tract's corners. A closure could be computed and the closed figure adjusted. This is the premise of the rectilinear or *tangent plane projection*.

The vast majority of individual real property parcels are mapped using the tangent plane projection system, as illustrated in Figure 4.2. The mapping is quite accurate provided that the area is small and centered on the meridian. As the distance from the base point increases, the relative positions of features on the Earth's curving surface are distorted on the flat map. The distortions caused by convergence of the meridians and curved surface distances are minimized by limiting the size of the area mapped. None of the state plane projection systems have adopted a tangent plane projection system because of the significant distortion that occurs once the distances mapped exceed 10 miles or so.

4.3.2 Lambert Projection

The *Lambert projection* system is best suited for mapping areas that have larger dimensions east-west than north-south (or, more simply, they are wider than they are long). The lines of unitary scale factor (SF = 1.0) are oriented in an east-west direction and the scale factors necessary to convert Earth surface dimensions to mapped (grid) dimensions can be held to less than the ideal maximum of 1:10,000 (meaning

---

[2]See section 4.1 of this text.

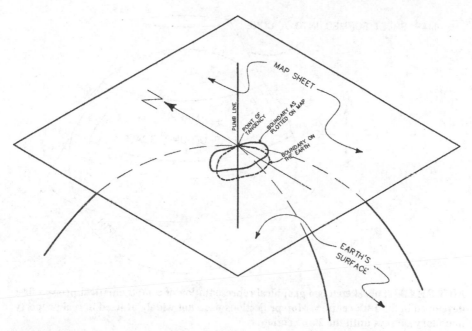

FIGURE 4.2 The most common projection system in use for small urban parcels is the tangent plane projection system.

scale factor is held to between 1.0001 and 0.9999) if the north-south dimension of the area mapped is less than 150 miles. Notice that the scale factor changes little for lines that run east and west but changes significantly for lines that run north and south. Lines of "unit scale" connect points where the scale factor is equal to 1.0, represented by the darker lines in Figure 4.3.

4.3.3 Transverse Mercator Projection

The *transverse Mercator projection* system is best suited for mapping areas that have larger north-south dimensions, than east-west (areas that are longer than they are wide). The axis of this cylindrical projection is perpendicular to the axis of the Earth. The radius of the cylinder is slightly smaller than the reference spheroid (*Clarke's Spheroid of 1866*, with a surface approximating sea level), resulting in two intersections between the cylinder and the spheroid with two parallel circles equally distant from the central meridian or base reference line. The lines of unitary scale factor (SF = 1.0, approximated by the darker lines defining the strip mapped in Figure 4.4) are oriented in a north-south direction, and the scale factors necessary to convert Earth surface dimensions to mapped (grid) dimensions can be held to less than the ideal maximum

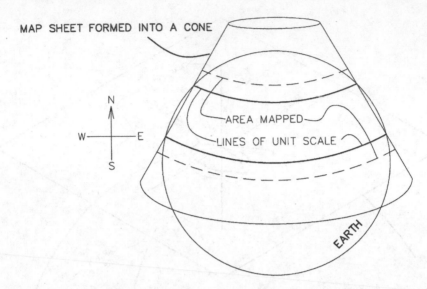

FIGURE 4.3 This sketch is a graphical representation of a mathematical process first developed in the 18th century. Map projections were not widely utilized in real property boundary surveys until the 20th century.

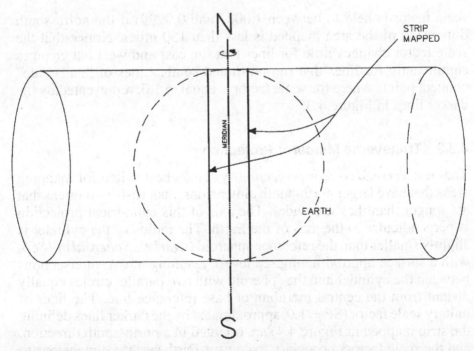

FIGURE 4.4 This sketch is a graphical representation of a mathematical process first developed in the 16th century.

of 1:10,000 (constraining scale factor to between 1.0001 and 0.9999) if the east-west dimension of the mapped area is less than 150 miles. Notice that the scale factor changes little for lines that run north and south while it changes significantly for lines that run east and west.

4.3.4 State Plane Projection

The governing authority of every state and territory has adopted a *state plane projection* system. The resulting *state plane coordinate system* (SPCS) in each jurisdiction is designed to meet the criterion of minimizing the distortion of distances throughout the area mapped.

SPC systems adjust for the curvature of the Earth by manipulating the distances between points. This manipulation is expressed by a term called the *scale factor* and is a function of location relative to the defined origin of the reference system or datum. The scale factor is a ratio reflecting the difference between a measured length and a length computed to fit the SPCS grid, correcting for the distortion resulting from projecting the ellipsoidal shape of the earth onto the flat surface of the state plane grid.

Distances are also defined by a vertical datum that can result in significant differences between surface measurements and distances along the datum. Localities of significant altitude differences relative to the geodetic datum will have measurable variations between points defined along the datum and distances along a level surface at that altitude. This second consideration is known as the *elevation factor.*

These two factors are often consolidated into a *combined factor* that all surface horizontal measurements must be multiplied by in order to compute state plane coordinate values for locations being mapped. Applying the combined scale factor to the measured surface or ground distance yields a *grid distance,* meaning a length that is consistent with the SPCS. Most states have broken the mapped areas into zones to limit the scale factor to between 1.0001 and 0.9999.

An example of a scale factor that is near the center of a mapped area might be 0.99995. That is to say, 1,000.00 feet measured on a level surface between two points on the boundary of a property parcel would compute as 999.95 feet between the SPCS values for those same two points. A scale factor for a property parcel at an altitude much below the SPCS datum might have an elevation factor of 1.00003; 1,000.00 feet measured between two points at such a low-lying property would compute as 1,000.03 feet between the SPCS values for the end points of that same line. The combined factor results from multiplying the scale factor times the elevation factor, in this instance (0.99995 $\times$ 1.00003) and equal to 0.99998.

Unless the area being surveyed is very large (over several square miles) or involves distances in excess of 20 miles, an average scale factor can be used as a constant for the entire project. High-precision work and surveys of large areas require a deeper investigation of the implications of scale factor and angular distortion. Those factors are beyond the scope of this book.

The original development of the concept of state plane coordinates during the 19th century did not anticipate the need to apply the scale factor to distances measured during ordinary surveying work. The precision of distance measurement by the boundary surveyor at that time was far inferior to the 1:10,000 error constraints introduced by the SPCS. Surveyors would traverse between control stations established by the governing authority. The distances measured during the traverse were adjusted using the Compass Rule or some other mechanism for distribution of errors. The distortion of scale due to the SPCS (as illustrated in Figure 4.5) was simply adjusted away. Today, the boundary surveyor is capable of a level of precision that was unimaginable during the development of SPCS.

The practice of ignoring the scale factor in an SPCS is surprisingly widespread, and often unwarranted. Most data collection systems can be set to automatically compute "ground to grid" conversions and vice versa. The vast majority of computations related to boundary determination, subdivision of tracts, resubdivision of PLSS sections and lots, and mapping are performed within computer-aided drafting programs that can also make these computations. The surveyor must

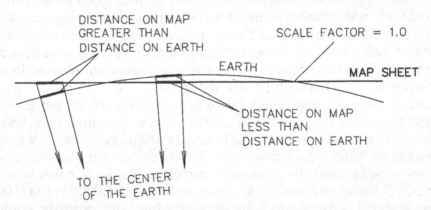

FIGURE 4.5 The surveyor has the option of presenting the distances on his or her survey plat in terms of level ground distances at the average elevation of the site or the grid distances between corners on the projection system. Survey plats presenting state plane coordinates should state if grid or ground distances are shown.

be aware of the projection system used in the mapping of any project and make conscious decisions about the need to consider and adjust for scale factors.

Survey plats that are mapped based on an SPCS must identify the specific SPCS adopted. Notes stipulating that directions are SPCS grid bearings and that distances are ground distances (or grid distances if that is the case) must appear on the plat. Descriptions of real property parcels mapped using an SPCS should also state that fact in the description.

4.3.5 Universal Traverse Mercator (UTM)

Worldwide mapping projects, such as the popular United States Geological Survey topographic maps, have adopted a *universal transverse Mercator* (UTM) grid system that divides the entire globe up into map strips of six degrees of longitude in width from pole to pole. The coordinates in this system are expressed in meters. The zones are numbered 1 to 60 eastward from the International Date Line.

An important characteristic of the UTM projection is that angles measured on the map or computed from coordinates on the grid conform closely to their true values. This is a transverse Mercator projection that is varied uniquely in each of the 60 UTM zones to minimize distortion to no greater than one part in 1,000 within each zone; each zone has its own distinct coordinate system. This means that multiple coordinate systems must be used when working with large areas, and those unaware of this fact will plot points representing areas in two different zones improperly with respect to their true physical relationship to each other. This can result in a more easterly property being mapped west of a more westerly property. UTM is a conformal projection, maintaining the shape of landforms while allowing relative size to vary on a map.

4.3.6 Global Positioning System (GPS)

The *global positioning system* (GPS) maintained and supported by the United States military is a navigational program, providing location and time information based on a constellation of satellites orbiting the globe. Unlike the previously described navigational and mapping controls, GPS, also referred to as a global navigation satellite system (GNSS), is an Earth-centered system. Increased positioning precision through use of GPS has resulted in a redefinition of coordinate systems throughout the world and has greatly reduced the imprecision and unknown distortions that plagued the surface-based system codified in 1927 (North American Datum, a polyconic projection). Utilizing

GPS to define a three-dimensional location (latitude, longitude, and altitude) provides a means of continuous refinement of mapping control parameters.

The modern land surveyor is no longer challenged by labor-intensive tasks merely to obtain reliable distances and directions. Real-time kinematic (RTK) applications of GPS technology allow the private-sector land surveyor to collect data and establish control precisely and over vast distances. Though currently not as precise as other data collection systems, such as electronic distance measuring equipment, RTK data is sufficient for most topographic data collection. RTK systems can provide real property boundary locations with sufficient precision to meet most state standards only if a rigorous and redundant observational routine is adopted.

4.3.7 Geographical Information Systems (GIS)

Geographical information systems (GIS) are rapidly replacing paper maps as a mechanism for presenting, documenting and preserving information that includes a location-dependent vector. Now that GPS has made determining location simple and universal, every aspect of life can be associated with a location.

The range of geodetic precision attained by the multitude of organizations collecting the information is vast, and unless documentation in the form of metadata is correctly attached to data and then researched by data users, many problems arise. The electronic methods used to retrieve the information do not take into account the level of precision of the method used to acquire or input the information, and the ability to zoom in on an image produces a false sense of its accuracy. As a result, unsophisticated users of a GIS can unwittingly introduce confusion and blunder into real property investigations.

For example, digitized United States Geological Survey topographic quadrangle sheets (USGS quad sheets) can be no more precise than the original map. Map standards in place during creation of the USGS 7.5-minute quad series allowed for a mislocation of about 60 feet between the graphically presented location of a feature and the true location of that feature. Many viewers of this data on a GIS do not realize this level of precision may not meet their needs when utilizing a USGS quad sheet as a base map, especially when the image has been enlarged on the screen beyond its original scale.

USGS quad sheets also, for general informational purposes, show PLSS township, range, and section lines in the states where they are applicable. These graphical lines were not field-located by surveyors or mapmakers. Instead, most of these PLSS lines were placed on the USGS quad maps by matching terrain features between PLSS field notes and aerial photography. The many years separating the origins of these two data sources introduce differences due to various survey methodologies, interpretation of field notes, and changes to physical features over time.

As recently as 2009, the US Army Corps of Engineers presented a map and "legal description" to be used in the acquisition of a particular right-of-way. The commencing point for the description was a section corner in a remote area that had never been surveyed (it was a projected section), yet the Corps map reported the location of that unrecovered corner to a precision of one ten-thousandths of a foot! Someone had clicked a cursor on the digitized USGS quad map and reported the SPCS values that the computer spit out as the location of that section corner, and to an extraordinarily high level of precision rarely (if ever) employed in writing land descriptions.

4.4   APPLICATION

The preceding technical discussions may have seemed far removed from the topic of property descriptions. But it is the responsibility of the professional surveyor to understand proper documentation of all the elements included in every description he or she prepares. It is also important for the surveyor to rely on an appropriate datum or reference system in conducting the field work and to know how to properly adjust measurements based upon the selected datum or projection. While it is not yet common for surveyors to lose credibility for relying on inaccurate or inappropriate coordinate systems, there have been lawsuits in which this has factored into the opinions of the courts. With the proliferation of GPS and GIS, it becomes increasingly important for the surveyor to verify data collected, analyzed, and manipulated by technology and software. Reporting coordinate values for corners in new subdivisions is a requirement for approval in more and more jurisdictions, but whether the reported coordinates are accurate is not always part of local government review and approval processes.

There are a number of ways to reference the system in which the survey was conducted and reported, some more detailed than others, some universally applicable, some varying with regional or agency requirements.

In the case of *California v. Arizona* (452 U.S. 431, Supreme Court of the United States, 1981), the state of California had initiated suit against both Arizona and the United States to quiet title to parts of the bed of the former channel of the Colorado River after the river had shifted position. The location of the river had particular importance to jurisdictional claims, as it forms the boundary between the two states. The United States was found to have no right, title, or interest beyond navigational rights. The Special Master appointed by the court submitted a report that both California and Arizona accepted, with portions of that final description reproduced here.

EXHIBIT A

A parcel of land in the former channel of the Colorado River in Imperial County, California, adjacent to Township 9 South, Range 21 East, San Bernardino Meridian; Township 10 South Range 21 East, San Bernardino Meridian; Township 10 South, Range 22 East, San Bernardino Meridian; Township 11 South, Range 22 East, San Bernardino Meridian, more particularly described as follows:

BEGINNING at a point on the center line of the former channel of the Colorado River having California Coordinate System, Zone 6, coordinates of x = 2,482,449.14 feet and y = 387,218.39 feet, from which United States Water and Power Resources Service (formerly United States Bureau of Reclamation) Station RUIN bears N 56 degrees 27'07'' E 733.37 feet, as said points are shown on the map entitled, "Davis Lake Area Project Administrative Maps," said map approved October 28, 1976 by the California State Lands Commission, and being on file at the office of said Commission in Sacramento, California; thence from said point of beginning, upstream along the center line of the former channel of the Colorado River, said center line being a fixed and limiting boundary of the herein described parcel, the following 377 courses: . . .

378. Northeasterly 278.84 feet along said westerly boundary, being a fixed and limiting boundary of the herein described parcel, on the arc of a curve, concave westerly, having a radius of 15,350 feet, to a point on said curve subtended by a chord which bears N 06° 51'25'' E 278.83 feet, said point being monumented with a standard California State Lands Commission brass tablet set in concrete, stamped "N-RB-Cal 1981," having California Coordinate System Zone 6 coordinates of x = 2,472,871.90 feet and y = 432,942.85 feet, from which California State Lands Commission Monument "PI-14" bears N 29° 11'26'' E 613.52 feet, as said monument is shown on said map; thence downstream along a fixed and limiting boundary of the herein described parcel, the following 27 courses: . . .

832. N 77°38'10'' W 213.25 feet to a point on the center line of the former channel of the Colorado River, said point also being the point of beginning of the herein described parcel of land.

Bearings and distances in the above description are based on the California Coordinate System, Zone 6.

While this Special Master's report included the basis for coordinates provided, the courts are not specific about the manner in which a report is to be prepared, leaving it to the expert to decide the best means of stating the location of the litigated lines. This can result in oversight of necessary references, such as in the Supplemental Decree issued by the Supreme Court of the United States in *United States v. Louisiana*.[3]

In that argument over natural resource rights in offshore waters, the court had been asked to establish "a baseline along the entire coast of the state of Louisiana from which the extent of the territorial waters under the jurisdiction of the State of Louisiana pursuant to the Submerged Lands Act can be measured."[4] But "Exhibit A" in this case differs dramatically from "Exhibit A" in the case just described. It launches directly into a table of coordinate values without stating the system in which they were developed beginning: *"A line from (X, Y), through (X, Y), through (X, Y)...."* and continuing in this manner for about 14 pages.

While the Special Master's report likely included additional references, the published court opinion is the document most publicly accessible, and anyone needing to rely on the court-established boundary must research further before being able to utilize the reported coordinate values. This is an example of what *not* to do in writing descriptions.

There are many means by which to provide a complete, thorough, and accurate report of boundary locations in a written description. The following is yet another variation of documenting a site's position in relation to a given reference system (this time, not the subject of litigation):

> ... Beginning at a buried railroad spike found marking the southeasterly corner of the herein described Lot 14.01, also being the northeasterly corner of Lot 14.06, being lands of Anthony Smith. Said buried spike is located in the traveled way of the public road known as Gully Run Road, (50 ft. wide right-of-way), and marks the same beginning point to these premises as recorded in Deed Book 2156, page 898, and has New Jersey State Plane Coordinates (NAD 83) North 599677.69, East 350614.69 (combined scale factor 0.99990117). From said spike and in said bearing base, running thence....

---

[3]422 U.S. 13, 1975.
[4]422 U.S. 13.

CHAPTER 5

PLATTING TO DESCRIBE

5.1 GENERAL

Ideally, every metes and bounds land description is based on a survey of the parcel being described. However, particularly in the non–Public Land Survey System (PLSS) states, we do suffer from the history of "protractions" (office drawings subdividing land into smaller lots that are then described based upon the plan of the protraction) and "partial surveys" in which only enough was surveyed to establish the boundaries of a new tract to be subdivided from a larger property, with the remaining land described but not surveyed.

But, for the most part, we can say that a metes and bounds description is based on a survey of the described tract, no matter the quality of that work. Ideally, an educated and experienced professional land surveyor produced that survey, but it could be the result of nothing more than a landowner's walking the perimeter of the parcel being created. While it is possible to create confusing and obscure descriptions from complete and concise surveys, poor descriptions are more commonly the product of less-than-professional land surveys. Indeed, incomplete surveys and the plats of those surveys can never result in an accurate, clear, and concise land description.

We have no control over the surveys performed in the past. But the modern professional land surveyor does have complete control over the work that he or she performs. Much of this book is directed toward understanding, interpreting, and clarifying descriptions of parcels that were created from obscure or erroneous surveys and survey plats. One of the goals of this book is to define a process that will result in metes and bounds land descriptions that are incapable of misinterpretation, even by parties that are determined to bend the original intent to their own personal purposes. The first step in producing such a land description is to perform a reliable and complete land boundary survey. Only then can a proper land description be penned.

Reliable land descriptions are clear, concise, unambiguous, and complete. Their purpose is to provide future landowners and other interested parties with the information necessary to find the physical limits of the described parcel on the ground. The inclusion of a metes and bounds description in an act of sale would not be necessary if all that was desired was a transfer of rights. The land description is necessary to clarify the limits of those rights on the ground, not just in the conveyance records. The land surveyor, who is required by state licensing statutes and regulations to understand real property rights as well as the means to measure location, is the professional best equipped to author reliable land descriptions.

The land boundary survey and the resulting survey plat need to present all the information required to identify the boundaries of that parcel in a way that cannot be misconstrued. Some of the essential items land boundary surveys must address to support land descriptions are:

- The identity of bounding parcels
- The natural and artificial boundaries of the parcel
- The orientation of the boundaries
- The dimensions of the boundaries

The language and parameters of the modern land boundary survey plat must be universally understood and clearly defined. Ideally, directions and distances will be expressed in terms that have nationally established definitions. Local lexicons should be avoided, or at least their meanings explained in more universal terms when such terms help to preserve evidence of original intent.

5.2   ORIGINAL SURVEYS

Landowners engage the services of professional land surveyors as their agents because of the confidence that the community has developed in them. Land surveyors are expected to have specialized knowledge in the location of the boundaries of real property parcels. The public counts on professional land surveyors to have the competence to measure and describe boundary lines and to know what must be done to ensure that the resulting land boundary descriptions conform to the intention of the landowners who have engaged their services. That intention must be followed during the course of a land boundary survey and then must be clearly expressed on the survey plat. Landowners will express their intentions in terms familiar to them. The land surveyor must understand those terms and then describe the land boundary locations and dimensions in a way that both the landowners and the world at large can understand.

5.2.1 Identifying the Bounding Parcels

A simple statement of distance and direction does not define a unique line, as there are an infinite number of lines that can have the same bearing and length. A real property boundary is a unique line in the universe. No two real property boundaries have exactly the same descriptive features: only one particular line can have a specific bearing, length, geodetic position, monumentation, and bounding parcels. Including references to parcels sharing a common boundary helps orient the reader of a description, whether those bounding parcels are identified by lot number, by deed citation, by owner, or other means. Further, such references, as in Figure 5.1, can help trace the chain of title in seeking the history of the creation of a parcel or confirming the pedigree of monumentation on a parcel's boundaries.

It should be remembered that, at times, the bounding parcel could be a right-of-way owned by a governmental entity, as when a county or state owns the bed of a road. In such instances, the current width of the right-of-way and the document establishing that width should be cited in the description for your tract. For some rights-of-way such as canals and railroad corridors, it can be difficult to find deeds or recorded plats confirming lines of ownership by private, public, or quasi-public entities. However, state archives and historical organizations sometimes house these important documents, providing significant evidence as to the intended boundaries.

The deed to a house purchased by one of the authors in 1980 provides an example of the importance of researching and referring to bounding parcels. At the time, the most the current description on file, written in 1968, was bounded on the west by the street, then "running easterly

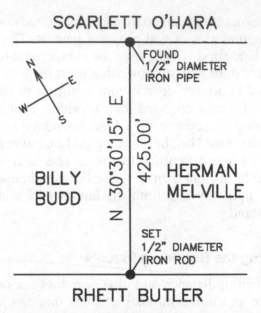

FIGURE 5.1 The reader may readily imagine that there are an infinite number of boundary lines in the world that are 425 feet long. Many such lines may also be along the bearing North 30 degrees 30 minutes 15 seconds East, some of which may also have their ends marked by half-inch diameters pipes or rods. But only one line could possibly divide a 19th century author from his fictional character, while beginning and ending at the boundaries of two Margaret Mitchell creations.

165 feet more or less along lands of Cadwallader, thence running southerly 25 feet more or less along lands of Anderson, and thence running westerly 165 feet more or less to the point of beginning."

In the process of researching public records to perform the survey of the soon-to-be purchased tract, which we will call "Parcel A," the 1968 description was found repeated verbatim through four earlier conveyances, back to 1926; it was not surprising that neither Cadwallader nor Anderson still owned the abutting properties in 1980. In the interim since 1926, evolving references in the adjoining deeds regarding the identity of the owner of "Parcel A" confirmed the changes in ownership found in "Parcel A's" own chain of title. From this corroborating evidence, the description in an adjoining deed of a relatively recent boundary line agreement executed along the northerly line of "Parcel A" provided a reliable basis for additional research to confirm acceptance of that agreement by an earlier owner of "Parcel A"—an agreement never referenced in deeds conveying "Parcel A" to new owners over the decades.

While it is hoped that such detective work and puzzle solving are not necessary when examining more recent descriptions, the fact remains that every additional piece of evidence provided may assist some future reader of the description in a way not anticipated. A deed that is unambiguous one year may prove less clear in the future, as evidence is lost and physical conditions and landowners change.

Let's look at an example of bounding references relating to a recorded plat of subdivision. The current deed of record contains three tracts, the first two reading as follows:

Tract No. 1

All that certain lot or piece of ground with the messuage or tenement thereon erected (being the larger part of Lot No. 213 on a certain plan of lots surveyed and laid out by J. Streeger August 28, 1890 for the Estate of Davis Jones, deceased).

Situate on the Northeasterly side of Price Street (40 feet wide) at the distance of 213 feet 10-1/4 inches Southeastwardly from the Southeasterly side of Jefferson Street.

Containing in front or breadth on the said Northeasterly side of Price Street 25 feet and extending in length or depth Northeastwardly of the width between parallel lines at right angles to said Price Street on 100 feet.

Tract No. 2

All that certain lot or piece of ground, being Lot No. 214 on a certain revised plan of lots of Davis Jones Estate, surveyed and laid out by Steeper and Zuschnitt, dated 5/30/1902 and recorded in the Office of the Recorder of Deeds at Norristown in Deed Book 493 page 500, said described agreeably thereto as follows, to wit:

Beginning at a point on the Northeasterly side of Price Street (40 feet wide) at the distance of 275 feet Northwestwardly from the Northwesterly side of a 30 foot wide street sometime called Rockland Avenue.

Containing in front on said side of Price Street 25 feet and extending Northeastwardly of that width in depth 108 feet 9-5/8 inches.

Bounded on the Northwest by Lot No. 213, on the Northeast by Lot No. 205, on the Southeast by Lot No. 215, and on the Southeast by Price Street.

Each of these descriptions contains a reference to a survey and subdivision. Each contains a width and angular relationship to the streets shown on the relevant plat. But then the comparison begins to differ.

Tract. No. 2 includes bounding calls along each of its sides. But Tract No. 1 cannot reference bounding lots because it is "the larger part of Lot No. 213," indicating a change in dimensions after the plat was recorded. Here is where a close examination of the referenced plat (to determine the point of beginning for Tract No. 1) must be combined with further record research, both of plats and of adjoining deeds.

Does the "revised plan of lots of Davis Jones Estate" referenced in Tract No. 2's description reflect the current configuration of Tract No. 1? Do any of the neighboring deeds reflect "the smaller part of Lot 213" in relationship to the original 1890 plan for the Estate of Davis Jones? If Tract No. 2 had not been included in the same deed as Tract No. 1 due to being part of a separate conveyance, merely reading Tract No. 1's description would not inform us of the existence of a revised plat. Simple recital of a lot in accordance with a recorded subdivision plat may not reveal the entire picture.

5.2.2 Monumentation

The natural and artificial monuments identified as marking the location of property corners are the physical objects that landowners rely on to find the limits of a real property parcel. Owners do not measure off a distance and bearing every time that they wish to exercise a real property right. The survey plat must identify the objects that memorialize the location of property corners. The identity must include the size, shape, material, and other features of the objects as well as the origin of the object. An iron rod found by a surveyor that can be identified as the mark set by the original surveyor of a parcel carries more weight than an iron rod found that has no history; objects or markers never referenced in descriptions have no historical basis as intended line or corner markers.

Claims to fences meant to keep cattle from wandering are the frequent subject of litigation by later landowners believing those fences mark property lines. Innumerable cases across the country underscore the point that fences unmentioned in deeds have no weight as evidence of intended boundary lines (although in some instances other evidence beyond the description may possibly alter this outcome).

Furthermore, markers unrelated to an original survey and not referenced by that original survey and description are unconvincing as "verification" and instead fall into the category of "assumption" of the location of boundaries. At the center of the arguments in *Dittrich v. Ubl* (13 N.W.2d 384, Supreme Court of Minnesota, 1944) was the presumed but unsubstantiated authority of an iron pipe in the northwest corner of the intersection of Minnesota Street and Seventh South Street in New Ulm, Minnesota known as the "Behnke corner" (so named for the person who lived at that street corner).

While surveyors had relied on this iron pipe for 20 or 30 years, the "Behnke corner" could not be traced back to the original monuments established during the first survey of the area on which the plat had been drafted and all the development in the area had been based. Instead, both Dittrich's and Ubl's deeds referenced the 1858 plat creating the city of New Ulm and showing a monument at the intersection of the centerlines of Broadway and Center streets. Relying on the "Behnke corner" would mean that Ubl's barn would be encroaching onto Dittrich's lot by 2.5 feet. Relying on the evidence presented by and related to the plat, the barn would be located only on Ubl's property.

The court noted:

> It is a well-settled principle that if lands are granted according to an official plat of the survey of such lands, the plat itself, with all its notes, lines, descriptions, and landmarks, *becomes as much a part of the grant or deed as if such descriptive features were written out upon the face of the deed or grant itself*, and controls so far as limits are concerned.[1] *[Emphasis added]*

The court went on to quote *American Jurisprudence* (a legal encyclopedia):

> Monuments set at the time of an original survey on the ground and *named or referred to in the plat* are the best evidence of the true line.[2] *[Emphasis added]*

The identification of every monument should be spelled out rather than assuming all readers will recognize the abbreviations so well understood by the author of the description. Perpetuate the pedigree of monumentation by stating whether it was found in the field or set as part of the current survey. Such information is invaluable to those following behind us as they try to recover the original surveyor's footsteps and not just the last footsteps on site. Figure 5.2 provides an example of how boundary evidence can be perpetuated.

Be specific in describing found monumentation; not all iron pipes look alike. They are made with different diameters and different thicknesses. Some are rusted, while others are shiny and new. The top may be crimped or split or twisted in a distinct manner recognizable as the work of a specific surveyor. Not all concrete monuments look alike,

---

[1] 13 N.W.2d 384 at 388.

[2] Ibid. It should be mentioned that the current monument located at the intersection of the centerlines of Broadway and Center streets was a replacement for the original and not the original monument itself. However, its location could be verified as being the same as the original marker through a variety of physical evidence discussed in the case.

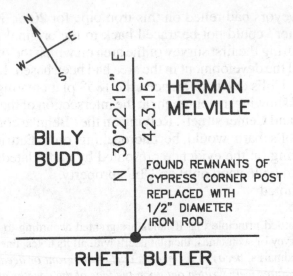

FIGURE 5.2   Consider the value that future generations will grant the half-inch iron rod found during a recovery survey when they can link that rod to the historical cypress post set by the original dividers of the tract.

either. They vary in size, and some may have chisel marks, while others may bear metal disks set into them.

And we should not always assume that the center of the monument is centered over the point in question. After encountering a disgruntled neighbor who did not want a client's corner monument on his land, one of the authors of this book found it necessary to set the face of the monument on line, with a deep chisel mark scribed on the side over the actual point. The resulting written description reflected the position of the monument in relation to the corner it marked, including specific mention and description of the chisel mark.

Furthermore, if there is more than one presumed "monument" in the area, the description should note the overabundance of evidence and clarify its relationship to the boundaries. Offsets to objects other than the intended corner marker can be identified and located in relation to the corner, and "passing calls" as lines cross over found markers on the way to the corner can be included with a distance from the origin or termination of the described course. See section 5.3.5 for examples of such references.

Such specificity in describing the intended monumentation guided the court in *Hubbard v. Dusy* (22 P. 214, Supreme Court of California, 1889) when finding not one but two lines of monuments (or "supposed monuments") along the line dividing Townships 14 and 15, Mount

Diablo base and meridian. Lacking better evidence, the state's Supreme Court affirmed the lower court's instructions that "other things being equal, the monuments, or line of monuments, most nearly conforming to the field-notes would be most likely to be, and would be considered to be, the true government corner or corners."[3] Both field notes and original survey plats derived from those notes contain evidence that must be reflected in descriptions based upon them.

Surveyors frequently find discrepancies between found monumentation and the written record. Sometimes a deed calls for monumentation, but what we find differs considerably in character or size from what is cited in the written record. Sometimes a deed calls for no physical evidence, but we find evidence in the field that somewhat "fits" the bearings and distances recited. Perhaps we find the evidence that is called for in the deed but discover that it has no relation to the original survey. The surveyor has a professional responsibility (to the client and the public at large, including future surveyors) to avoid perpetuating such ambiguities, a responsibility that can be fulfilled by including clear and complete descriptions of monumentation found and set in every written property description.

But it isn't only the usual physical corner markers that can be confusing. In the instance of *Hopkins v. Black* (310 P.2d 702, District Court of Appeal, 3rd Appellate District, California, 1957), the called-for monument in question was "Flea Valley Road." However, there was evidence of two such roads in the vicinity that could fulfill the deed's reference to "the center of the County Road from Flea Valley to Concow" as the easterly and southerly boundaries of the tract, and the court had to determine which was the road intended by the original deed. Road Number 1 would create a tract of approximately three acres, which would closely agree with the area called for by the deed. But relying on the current location of Flea Valley Road, Road Number 2, would create a tract of about 9.2 acres, a discrepancy of approximately 6.2 acres.

The court heard testimony from various witnesses as to where each road ran (between what towns), the compass direction each road traveled, and their locally known names. It then determined, in conjunction with oral testimony, that the direction of the road called for in the deed (southwest) overruled the call for acreage, and that the present location of Flea Valley Road, Road Number 2, was the one intended by the deed.

Surely, it is impossible for the drafter of a description to foresee future roads in the vicinity of the site being described, and old road names can be confusing. But *Hopkins v. Black* does not present a unique

---

[3]22 P. 214 at 215.

situation. If ever traveling in Camden County, New Jersey, be careful when asking for directions to any destination on "Erial Road." The answer will be, "Which one?" In this long-settled agricultural area, the following are all possible locations, any of which may be referred to as "Erial Road" for brevity:

- Old Erial Road
- New Brooklyn Erial Road (sometimes referred to as Erial New Brooklyn Road)
- Erial Clementon Road
- Williamstown Erial Road (and sometimes Erial Williamstown Road)

Being forewarned about such potential confusion due to similarity in names and proximity of the roads to one another, description writers should include additional information that makes the intended location unique. This can include local and formal names for the road, the right-of-way width, abutting owners, and deed references to pin down locations by evaluating time, that often disregarded fourth dimension.

5.2.3   Directions

There are a multitude of methods of describing the orientation of a land boundary. Each method has advantages and disadvantages.[4] The professional community surrounding real land transfers has begun to settle on the north-based bearing as the preferred method of reporting the orientation of a boundary. However, the basis of that orientation must be referenced in the description.

Reference to the basis of directions can resolve some boundary disputes, and the case of *Scott v. Yard* (18 A. 359, Court of Chancery of New Jersey, 1889) shows us the importance of historical orientation (Figure 5.3).

Scott owned three tracts in Wall Township, New Jersey. These lots were described in his deed as each being 50 feet wide by 150 feet deep, fronting on Ocean Avenue, and facing the Atlantic Ocean, which was directly on the other side of the Avenue. His deed, dated August 26, 1880, was part of a larger tract that had been platted, subdivided, and filed in the County Clerk's office just 15 days earlier.

---

[4]See Chapter 3 (and particularly section 3.3 regarding bearings) for more detailed discussion.

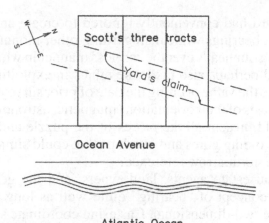

Ocean Avenue

Atlantic Ocean

FIGURE 5.3 *Scott v. Yard.*

But Yard claimed a 45-foot wide strip through Scott's three lots, splitting the front of the lots from the rear portions, thereby cutting the rear of Scott's lots off from access to the road and making his lots unusable. Yard refused to sell his claim to Scott, except at a cost well in excess of the true value. Scott filed suit to dispute Yard's claim and make his own lots whole again.

Scott based his claim to title on two surveys conducted more than 100 years apart; the west part came from a survey and plat for Forman dated 1746, and the east part came from a survey and grant to Brinley dated 1860. The decisive question facing the court was if these two tracts were on a coincident line as Scott argued or separated by a strip of land 40 to 45 feet wide as Yard claimed—based on a survey Yard had made himself as deputy surveyor under the Board of Proprietors of East Jersey to find unappropriated and unpatented lands outside of prior grants, and dated October 19, 1880, about two months after Scott took title to his platted lots.

The court heard testimony from the surveyor of the Brinley tract, who stated that he had left no gap between his 1860 survey and the earlier 1746 survey. (Yard himself conceded that Scott's lots were part of the 1860 and 1746 surveys, despite claiming an intervening strip between those surveys.) The court also examined the points of beginning for the various deeds involved in the dispute, and compared the bearings of the lines coming from the common beginning point to a report of the United States Coast Survey regarding magnetic declination. In doing so, it found that the difference in bearings between the 1860 and 1746 surveys coincided with the change in declination over the years, a change that Yard had conveniently ignored to create an artificial gap based solely on bearings while disregarding other monumentation.

After shaking our heads over the careless manner in which Yard made his survey (and perhaps also his loose ethics in exploiting his official post), we realize the value of citing the date of prior surveys on which we relied and the basis of their orientation (magnetic, astronomic, assumed, etc.). We should present all the pieces of the puzzle and explain their significance, avoiding gores and overlaps that could simply be due to a variation in the north arrow reference.

Lawyers, real estate agents, landowners, and the general population grasp the concept of "bearing" quite well as long as we restrict ourselves to the two-dimensional, Cartesian coordinate system.[5] Modern advances in the science of geodetic positioning have narrowed the range of acceptable systems of describing directions even more.

Precise geodetic positioning is so inexpensive and universally available that there is simply no excuse for not defining directions using the state plane coordinate system. State plane coordinate (SPC) systems are two-dimensional, flat-plane systems readily accepted by the public. Lines of constant bearing are straight. Parallel lines have the same bearing, and the bearing of a straight line is the same coming and going. The SPC system is two-dimensional and consistent with the flat-earth intuition of us all. State plane directions, when defined as bearings, are developed by a system of observations that are not dependent on the specific monumentation of the parcel being surveyed.

Even prior to the development of global positioning systems (GPSs), a surveyor could determine SPC directions from solar or stellar observation. The process required some effort. The location of the observation station had to be determined to within one second of arc, latitude, and longitude. This was usually determined by scaling these values from a United States Geological Survey (USGS) topographic map. The hour-angle method also required a very precise recordation of the time of observations. This was usually accomplished by tuning in to radio stations specifically dedicated to broadcasting universal coordinated time.

Sometimes the radio stations could not be received. Sometimes the terrain was such that determination of latitude and longitude by scaling was not possible. All computations were completed by hand or with the assistance of a calculator capable of only basic arithmetic functions. Few surveyors believed that the extra effort of solar or stellar observations was worth the resulting precision in directions.

---

[5]See Chapter 4 for a more comprehensive discussion of coordinate systems.

Today, most of the difficulties associated with precise solar determination of direction have been solved. Solar observations require the observer's location to within 60 feet or so and coordinated universal time to within a second. Both are available in the civilian GPS receiver. A solar observation requires about 15 minutes of additional field time. Programs that reduce the field observations to SPC directions are readily available. Most survey-supporting computer-aided drafting (CAD) programs include solar and stellar routines as part of the standard software. Survey-grade GPS receivers and software routinely produce positional results defined in SPC parameters.

Modern survey hardware and software focus on developing data in terms of the national geodetic datum. The simplest, most direct and widely supported method of defining directions today is the SPC bearing. Survey plats published using this system support clear and concise land descriptions. However, no matter how precise the metes of a description, the intent of the parties and the physical evidence of bounds should never be set aside in favor of mere coordinate values. Doing so strips any illusion of accuracy from the process and demeans the value of real property rights to mere mathematical expressions. When coordinate values are reported, the writer of a description should make clear that these numbers are subservient and merely supplemental to other evidence in the document, and the fact of convenience in finding a position by its coordinates does not equate to an accurate location of real property interests on the ground.

5.2.4 Distances

In reading older deeds, the history of surveying shows us that there is sometimes a local difference in how distances were measured, with horizontal measurements not always being the norm. Depending on the terrain, it was not unusual for surveyors to lay the chain on the ground, producing a measurement between known monuments that was longer than what a horizontal measurement would yield. This practice was sometimes employed to simplify life for the surveyor in the field, but sometimes to assure that a parcel being divided from a larger tract would include adequate arable lands.

In other regions, horizontal measurements were employed, but differences between the lengths of various surveyors' chains could result in highly localized standards of length—still employed—that differ from the standard U.S. survey foot. In such instances, an outsider unfamiliar with "how to measure" can do as much damage as any surveyor with an uncalibrated tape or electronic distance measuring (EDM) device.

Similarly, a surveyor employing coordinates must be well informed about corrections and transformations before introducing SPC values into a description. When used properly, an SPC system provides certain advantages to reported position and direction, such as consistency and reliability. But the distortion of distances between ground and grid measurements, although small, must be identified and addressed.

Conversions between geodetic and plane coordinates and between metric and foot measurements are common, but must be undertaken with care. The foot is the standard unit of measurement for real property boundaries for most of the United States (historical local measurement units still do exist). There are two definitions of the foot recognized in the United States. The international foot is defined as 3.28084 feet to the meter, while the U.S. survey foot is defined as 3,937/1,200 feet to the meter. The U.S. survey foot is the definition used in all SPC systems. The difference between measurements made in differently defined units can be quite large. For example, an SPC location of y = 253,567.052 meters, x = 1,524,897.563 meters will properly translate to y = 831,911.24 feet, x = 5,002,934.76 feet in terms of the U.S. survey foot. But that same location would be identified as y = 831,912.93 feet, x = 5,002,944.92 feet if the definition of the international foot were mistakenly applied, a difference of 10.30 feet.

The advantages of utilizing SPC systems in locating and annotating real property positions are such that most municipalities require that surveys involving public recordation be referenced to a government-defined SPCS. How distances are addressed is problematic. Real property distances are local, level distances. The acreage sold and possessed is tabulated as local, level values. The surveyor must be aware of the consequences of mixing horizontal surface distances and grid or mapping distances.

Prior to the development of EDM equipment, the primary means of establishing SPC control for a survey project was to traverse, by chain and transit, between two published SPC stations. The closure errors for these traverses were much greater than the combined scale and elevation factors for all but the most unique projects. The act of "balancing the traverse" spread the measurement closure and scale factor adjustments throughout the traverse. Local measurements made from adjusted traverse stations were comparatively short in relation to the entire traverse, and so the differences between level ground distances and the associated theoretical grid distances for the real property boundaries were minute when compared to the standard deviations of the overall measurement process. Scale factors could be, and often were, ignored in this early work.

Modern measurement precisions and expectations are such that these differences cannot be ignored for the majority of real property boundary surveys. If the surveyor does not identify and control the scale factor aspect of SPC work, confusion can result. The equipment chosen to collect survey data can be adjusted to address scale factor, but it is the surveyor who is responsible for the consistent and appropriate interpretation of the data collected.

5.3 RETRACEMENT SURVEYS

Retracement surveys are not limited to the data presented on an original survey. A surveyor performing a retracement will often find boundary data that developed after the original survey or information that was gathered during fieldwork but not presented on the original drawing. But none of this evidence is presented in metes and bounds descriptions in which the preparer has pursued brevity rather than perpetuation of the field and record evidence that resulted in the final survey report. This commonly results from misinterpretation of survey standards (including American Land Title Association or ALTA surveys) that require the "legal description" to appear on the face of the survey plat. As we have discussed previously, a "legal description" is not necessarily a metes and bounds description.

A metes and bounds description that is written for parcels created under the PLSS or platted subdivision land record system can only supplement the description of the parcel, not supplant it. One of the authors was presented a deed to a regular PLSS parcel. The legal description of the parcel read as follows[6]: *the SE 1/4 of the NW 1/4 of Section 23, T1N R6W Vernon Parish Louisiana*. The following metes and bounds description was added to the deed after the phrase "more particularly described as":

> Beginning at the center of Section 23; thence, go West, 1320′; thence, go North 1320′; thence, go East 1320′; thence, go South 1320′ to the point of beginning, and containing 40 acres.

It is clear, in this case, that a recovery of the boundaries of the parcel was not performed or relied upon to write the metes and bounds description. This particular metes and bounds description does not clarify the true boundaries of the parcel; instead, it obscures the location

---

[6]Some details have been changed to ensure confidentiality.

of the corners and misleads the reader. The aliquot division of Section 23 could not possibly have resulted in an ideal quarter-quarter parcel.

The modern survey of the parcel recovered several corners (perpetuated) and a proper aliquot division of the PLSS section was performed. This resulted in a parcel that was slightly larger than 40 acres (40.25), yet the title insurance company refused to insure anything over 40 acres or any lands that fell outside of the 1320′ square, in an effort to maintain the metes and bounds description on record.

The modern surveyor should be aware of and avoid such failings. There is no legal requirement that the new metes and bounds description of a retracement survey must simply be a repeat of the old, and such practice thwarts preservation of the true boundaries. The modern surveyor performing a retracement will report bounders, markers, features, and other pertinent land information that was not present during the original survey. The modern survey may include a new base meridian so that the bearings are of different values than the original, in which case an explanation of the conversion should be included in the survey report. The retracement may be reported in feet instead of rods, chains, varas, arpents, or some other archaic dimensional unit.

The same holds true for the metes and bounds land description. The surveyor is free to include new information or delete erroneous, superfluous, or misleading words or phrases from the old survey as long as the action improves the readability and communication of the description and does not alter the intent of the original parties in creating the parcel or interests conveyed. *Real property boundaries are defined by the location of the corners.* Metes and bounds descriptions are intended to assist in the recovery of the corners. These descriptions do not, and cannot, alter the location of the corners.

However, merely transforming the older units into feet by applying a mathematical conversion factor erases evidence of the original survey and the methods used to accomplish it. Such conversion also frequently introduces new sources of confusion and errors by changing the significant figures of the reported lengths of lines.

As a very simple example, we will assume a rectangular tract of land reported as 2 chains 47 links wide and 5 chains long. Determining the area of such a parcel is a simple matter of multiplication: we know that there are 100 links in a chain, and that a chain (at least the common Gunter's chain) is 66 feet in length. To calculate area, we multiply 2.47 chains by 5 chains, and divide this product by 10. Thus, the area of this parcel is 1.235 acre. This figure is consistent with the method of original measurement, as chains and links are the historical basis for calculating acres.

Someone wishing to "modernize" this tract's description by converting its chains and links into feet is likely to multiply each reported chained length by 66 feet. In our example, 2 chains and 47 links can be thought of as 2.47 chains and then multiplied by 66, yielding 163.02 feet for this length. The original measurement was never made to the hundredth of a foot; it was made to the nearest whole link, or one one-hundredth of 66 feet. Anyone reading "163.02 feet" will accord that length much more precision than it deserves, expecting it to be a quarter of an inch longer than 163 feet. But the measurement had originally been made in chains *and links,* meaning that the significant figures should be related to the nearest whole link (approximately eight inches) rather than to the nearest one one-hundredth of a foot.

Similarly, our 5 chains for the length of our tract translate into 330 feet (5 times 66). Note that this figure contains no decimal places, meaning that the significant figures are to the nearest whole foot. This is not completely correct, either, as original measurements were made to a finer increment. There is also the possible misperception that significant figures in the reported "330 feet" are to the nearest 10 feet, simply because of the coincidental location of a zero in its final place.

All of this is to say that mere transformation of measurements reported in archaic units to feet or meters can obliterate the original intent and make it more difficult to discern possible sources of mathematical errors. When a description is reported in rods and it mathematically miscloses by a multiple of 16.5 feet, we have an idea that the first measurers miscounted the rods. But when that same description is converted to feet and hundredths of a foot, we lose all sense of how the land was measured and what the sources of errors might be.

5.3.1 Hierarchy of Calls

The hierarchy of calls, so important to boundary recovery, is equally important to survey plats and land descriptions based on those surveys. When drafting the plat, the hierarchy influences the surveyor's decision as to what should appear and what need not be shown.

The intention of the parties is the primary and overriding factor in applying the hierarchy. Rarely is the intention of the parties expressed as a phrase or a paragraph on the plat, although such phrases can often appear in a land description. The intention of the parties is most often expressed on the plat of survey when considered *in globo* ("as a unit" or "as a whole"). This topic is addressed in much more detail in section 6.2.

5.3.2   Identification of Lines

All boundary lines are common to some other landowner. That owner might be a private person or local, state, or federal government. There are no boundaries enclosing one parcel that do not exclude another. Riparian boundaries separate federal or state ownership from private title. Parcels defined by lot identification, aliquot divisions, sections, metes and bounds, or any other land record system consist of boundary lines that are contiguous with other parcels. Identification of the bounding parcels is fundamental to expressing the intention of the vested parties, and so survey plats and land descriptions must report those contiguous parcels.

Natural boundaries or monuments must be presented on the plat of survey when they occur. Rivers or rock outcroppings are hard to miss. It is vital that these features appear on the plat of survey and in the land description based on that plat. The fact that a major river bounds a property is something worth noting. While reference to the plat that includes such natural boundaries legally incorporates the plat into the description, it is far better to specify those physical features as boundaries in the written description as the intended limits of interest than to leave the door open to future argument about the intent of the description. Reference to the center of a watercourse as opposed to the high or low water mark includes full rights both to access the water and to own the submerged lands. But if a description limits conveyance to the high water mark, the purchaser does not have access to the water's edge at all times (except in common with the general public for whom the government reserves easement rights between high and low water). Lack of any reference to the water at all may, years later, result in a fixed boundary rather than one that moves with the gradual change in the watercourse location.

Artificial monuments are important to the recovery of real property boundaries. But while they rank above directions and distances in the hierarchy of calls, many survey plats and descriptions make no mention of them. Past failure to report such important information pertaining to the location of boundary corners has created conditions under which boundary recovery is often difficult and uncertain. Modern survey standards require the reporting of all artificial monuments, both found and set. Land descriptions written from these modern survey plats must also report the location and origin of artificial monuments used to mark the boundary corners.

Directions should be presented on the survey plat in terms most easily understood by the users of that information. Bearings (rather than azimuths or other directional references) have developed into the most common and widely accepted method of documenting directions for real property boundaries. SPC-based bearings have the added benefit of being supported by a universal geodetic system, as they can be recreated and recovered without benefit of a single physical monument reported in the plat of survey. Land descriptions that report directions in terms of the SPCS are more recoverable than any other method of documenting directions and provide a means of beginning the search for physical markers controlling the location, size, and shape of a parcel. As an increasing number of local, state, and federal agencies require geodetic orientation, survey plats that are not based on SPC systems are becoming rarer commodities. In terms of providing a good starting point for a search for property corners, coordinates (state plane or otherwise) are excellent tools. But it must always be remembered that coordinates themselves are a means of last resort in determining real property corner locations and cannot be the sole means used to define them.

Lay users of real property information most readily understand distances that are expressed in local ground surface, level feet. If grid distances are used, the survey plat and the land description must document and report that fact. Ancient and archaic standards of distance measurements should be reported on survey plats only if mentioned in the chain of title, and then accompanied by the U.S. survey foot equivalent. The use of the international meter in land records has not yet been accepted by the American public outside of Puerto Rico (where the metric system is employed in measurements), despite passage of several congressional acts attempting to move our country to the standard used by most of the rest of the world. Metric dimensions reported on plats and in descriptions should be accompanied by the U.S. survey foot equivalent.

5.3.3 Area and Significant Figures

Area is not a directly measured quantity. It is a computed quantity based on measured distances and directions, and therefore because it is derived rather than directly observed, it rarely defines the intent of the parties. Area is computed in terms of the measurement units employed during the survey, so the factors controlling distances also apply to areas. Square feet and acres are the acceptable standard reporting units for U.S. survey foot surveys. Similarly, ancient and archaic terms of area should be reported on survey plats only if mentioned in the chain of title and accompanied by the U.S. survey foot equivalent. The use of the hectare in land records in the United States has not yet been accepted by the general public, and may never be (except, of course, in Puerto Rico). Hectare values (in areas outside of Puerto Rico) should be accompanied by the U.S. survey foot equivalent in acres.

We introduce a considerable "round-off" variable when converting between acres and square feet. Therefore, highly valued land should be reported in terms of square feet as well as acres, and the surveyor should take care that the significant figures are consistent. The concept of significant figures is too often forgotten in the mindless repetition of the same number of digits spit out by our calculators and computers. However, "significant figures" relates to the particular digits carrying meaning related to precision, including all digits *except* for leading and trailing zeroes that are merely serving as placeholders, *or* digits that are merely reported due to calculations carried out to greater accuracy than the original data, *or* reported to a precision greater than the capability of the measurement equipment or technique utilized.

One acre contains 43,560 square feet, and this is directly traceable to the original tools used to measure distances from which area is computed. When measuring in chains (and not converting back to feet), area in acres is easily determined by multiplying the distances along two sides of a rectangular tract, and then dividing that result by 10. If we have a tract that is 3 chains wide and 2 chains long, the parcel contains 0.6 acre (3 chains times 2 chains, divided by 10). There are no decimals associated with this area beyond the result of division by 10, as there are no decimals associated with the chained measurements. A division of a Gunter's chain is not in decimals, but in links, each link being one one-hundredth of the chain's full length.

The same is true of measurements made in rods or perches (a rod or perch in the United States most generally being 16.5 feet long, although local variations do exist, ranging from 10 feet to 24 feet, due to a variety of historical European origins and influences). To the uninitiated, there can be some confusion in understanding areas since the term *perch* can represent either linear or area measurement. An area of one perch (not always referred to as one square perch) is equal to a square rod, and 160 (square) perches are equivalent to an acre. As when working in chains, if we keep our computations all within the same system and do not convert to feet or meters, we maintain the same relationship to significant figures as intended by the original parties to a deed, who would never have imagined reporting area to three or more decimal places.

When measuring in feet, however, we do directly measure to the hundredth of a foot when using a tape and possibly to some finer increment when using an electronic distance-measuring device. But we must be sure that the linear measurements our equipment displays are indeed accurate to the number of decimal places reported. If our techniques (or land conditions) do not support finer divisions, then a hundredth of a foot may be all that is accurate, although the instrument may be reporting to four decimal places. Have we calibrated the instrument; have we made the correct parts per million (ppm) adjustments for the reflectors we are using; are we measuring near an object that might cause refraction of the measured line? Similar questions must be addressed when measuring in meters, in order to ensure that the number of recorded decimal places have meaning and are accurate.

While the general mathematical rule is that multiplication of numbers containing decimal places set the number of decimal places that can be calculated, this is unrelated to the precision of the final result. The mere fact that we can measure one side of a rectangle as 99.54 feet and another as 100.47 feet does not mean that we should report the final result as 10,000.7838 square feet. The number of significant figures here includes only two decimal places, as we have not measured precisely enough to rely on the third and fourth decimal places of our calculation.

When we convert these square feet to acres, we divide by 43,560 square feet, a figure with no decimal places due to its historical origins as a derivation of chained measurements. However, when we divide our calculated area from our example by 43,560 square feet, we derive 0.2296 of an acre. Rounding this figure to fewer decimal places will result in an incorrect calculation of square feet when the rounded area in acres is multiplied by 43,560. So in this instance reporting four decimal places is reasonable.

This is not always true, however. A tract bounded by a body of water cannot have measurements along that volatile boundary to the nearest hundredth of a foot (unless a meander line or tie lines are calculated). Therefore, the plus or minus measurement of water boundaries means that the area in acres should be reported to only one decimal place since the calculated number of square feet is only a "more or less" figure.

The units of measurement reported in a description should be consistent, but not at the expense of losing the original intent when "updating" to a new system of units. Consider the following example, on file in the County Recorder's office in 2010:

> Beginning at a stake in the middle of Warner Avenue at the distance of 285.57 feet from a point in the middle of the Philadelphia and Lancaster Turnpike Road, thence along the middle of said Warner Avenue South 48 degrees 30 minutes

West 50 feet to a point, a corner of land now or late of John T. Harper; thence by said land of Harper South 41 degrees 30 minutes East 102.4 feet to a stake in the land now or late of Mary Murray; thence by said land of Murray North 48 degrees 40 minutes East 50 feet to a stake, a corner of land now or late of Lenia Stillwagon; and then by said land of Stillwagon North 41 degrees 30 minutes West 102.4 feet to the place of Beginning. Containing 18.8 square perches of land be the same more or less.

A number of discrepancies fly out at the reader, particularly the opening tie distance to the hundredth of a foot and the area reported in square perches. From the closing line we can surmise that this tract was originally measured in rods or perches. But such measurements would not be reported to the fine-tuning of the nearest one one-hundredth of a foot, as is presented in the tie distance from the intersection of two roads to the beginning point; unless this is an addition to the original description, the line's measurement should be to the nearest rod or perch, a unit equivalent to approximately 16.5 feet.

Square perches are calculated on the basis of 160 square perches comprising an acre. From this we can compute the acres of this tract:

$$18.8 \text{ square perches} \times [\text{one acre}/160 \text{ square perches}] = 0.1175 \text{ acre}$$

If this area is reported in terms of "acres," the temptation is to immediately transform it into square feet, which we could obtain through multiplying 0.1175 by 43,560 (square feet per acre). The figure thus derived, 51,118.3 square feet, is completely inconsistent with a measurement originally reported in terms of perches and implies a precision not possible when measuring in multiples of 16.5 feet.

5.3.4   Recovery of Monumentation

Surveys that recover monumentation set by previous work, especially work by the original surveyor, should document that recovery on the survey plat. Information pertinent to the location of real property boundaries presented on survey plats should be reflected in the land description of that parcel. The survey plat does not necessarily have to be restricted to only monuments found on the boundary of the parcel being surveyed. Monuments or features in the vicinity of the parcel that have a known geographical relationship to the subject property and that are relevant to an improved understanding of the real property boundaries (i.e., corners and lines) should be reported on the survey plat and in the land description if it will clarify the boundary location. This includes recovered markers of original tract lines from which the current surveyed parcel was divided, monumentation of street or right-of-way alignment, and other controlling features.

5.3.5 Perpetuation of Monumentation

Surveyors have a professional obligation to improve the monumentation of the original surveyor if the original marks are subject to deterioration. An original stone, for example, can be improved as a monument by the addition of witness pipes or rods to enhance electronic detection. An original pit and mound can be improved as monumentation with the addition of a pipe in the bottom of the pit. A pipe that is rusting away can be improved by the addition of an iron rod in the center of the pipe. A corner recovered by witness marks can be improved by the placement of a new marker at the corner.

All of these perpetuations of real property corners must be documented on the survey plat and included in the land description for the benefit of future recovery and proof of the corner. A full report of the material, the size, and the condition of both the original marker and the supplemental marker will underscore the purpose of the newly set marker as ancillary but not a replacement.

The relative position of the two markers to each other is a critical piece of information for landowners and future surveyors. While offsets from the actual corner to found markers in cardinal directions may prove useful to surveyors, however, the layperson has no real understanding of which way is north or east of a found marker to understand where the true corner is. Therefore, for markers that are not actually on a property line, often a bearing and distance between the marker and the corner provides solid information both for a surveyor and for a layperson who is likely to use "swing" ties to try to find the desired point.

If markers are found on the property line but not on the corner, reporting a distance along that line to the corner is sufficient. This is a common situation when the line terminates in a water body and the marker had to be set back from the shoreline. Such a condition can be described as follows:

> Along the southeasterly line of Lot 1, Block 6, North ten degrees fifty-seven minutes twenty-nine seconds East, two hundred seventy-seven feet and forty-six hundredths of a foot to a point in the Millstone River, passing through a boulder with cross-cut found one hundred and fifty-six feet and seven hundredths of a foot from the origin of this course; thence. . . .

The situation need not always be a watery one. Offsets from unstable soils, steep slopes, or traveled ways may also be required for more permanent marking.

Along the said easterly line of Williamstown-New Freedom Road, and passing over a concrete monument found 50.00 feet from the terminus of this course, North 30 degrees 43 minutes 43 seconds East a total distance of 439.71 feet to a point in the northerly line of Winchester Road (33 feet wide); thence. . . .

This kind of reference is often referred to as a *passing call*, since the described line passes over it at some point. Multiple markers of significance to the understanding of the line's location might be found and reported in this way. Whether including one or multiple passing calls, always take care to note if the reported tie distance is to the origin or terminus of the described course.

5.4 PRESERVING THE EVIDENCE IN WORDS: A CASE STUDY

Three very different kinds of descriptions surfaced while researching the chain of title for a New Jersey tract and its adjoiners. They illustrate a surveyor's eye to evidence of location, evidence of intent, and the difference between clarity and brevity.

The most current deed of record read as follows:

Deed 1 (dated 1978)

ALL THAT CERTAIN described land located in; Township of Kingwood, County of Hunterdon, and State of New Jersey more particularly described as follows:

BEGINNING at a point for a corner Northerly of a road known locally as Stompf Tavern Road, said corner being the Northeast corner to lands of Carl Woronowicz; thence

North 57 degrees 16 minutes 45 seconds East a distance of 220.00 feet to a point for a corner; thence

South 28 degrees 25 minutes 20 seconds East a distance of 447.35 feet passing over Stompf Tavern Road and running along lands remaining to Robert Chabot to a point for a corner; thence

South 57 degrees 16 minutes 45 seconds West a distance of 220.00 feet along Chabot to a point for a corner in a line of lands of the aforementioned Carl Woronowicz; thence

North 28 degrees 25 minutes 20 seconds West a distance of 447.35 feet along Woronowicz and passing over Stompf Tavern Road to the point and place of beginning.

CONTAINING 2.259 acres of land. . . .

While this deed closes mathematically, it is not very helpful in terms of determining its location on the ground. Most obvious is that each course merely runs "to a point," with little in the way of physical clues. Looking more closely, other questions arise.

We begin at a point "northerly of a road": How far northerly? Are we northerly of the physically traveled way or of the recorded right-of-way? What is the right-of-way width at the time of this recording? This road is "known locally as Stompf Tavern Road": Does it have other more generally known names? How local is local? We are at the "Northeast corner to lands of Carl Woronowicz," but Mr. Woronowicz is quite the land speculator: to which of his many tracts are we referred?

Course 1 gives no clue as to ties to adjoiners or relation to roads, monumentation, etc. Course 2 passes over Stompf Tavern Road, but we do not know at what point in relation to the beginning or terminus of this course such an intersection occurs. Fortunately we have a call to lands remaining to Robert Chabot, but we will need to find his deed to find out where his property lies as well. Course 4 terminates in a line of lands of our land speculator, Carl Woronowicz, so we are not certain which of his lands we have run into. (Although not at all certain or evident from this deed description's language, we have in fact been fortunate to encounter the same parcel at which we began.) Course 4 again does not specify at what point the course crosses Stompf Tavern Road.

As a last comment on Deed 1, note that all numbers are written in numeric form only, without being spelled out in letters. The possibility for transposition or outright blunder is more common with numerals. Rules of construction relating to ambiguities in deeds give precedence to written words over numerals when a discrepancy exists.

Because this deed did not provide much in the way of assistance, it was necessary to trace the title backwards to look for any possible clues regarding this tract's location and boundaries. In this process, we encounter:

Deed 2 (dated 1929)

. . . lying and being in the Townships of Delaware & Kingwood, in the County of Hunterdon and State of New Jersey, bounded and described as follows, viz:

Lot No. 3-

********** BEGINNING at the southeasterly corner of Lot No. 2; thence (1) along Lot No. 2, North thirty four degrees and ten minutes east crossing the public road and passing on the southerly side of a blazed Ash tree on the easterly side of said road Five Hundred and fifty nine feet more or less to a corner; thence (2) South forty six degrees East one chain and fifty six links to a White Oak tree; thence (3) North eighty eight degrees forty five minutes

East four chains and twenty five links to a stone heap on top of hill in line of land now or formerly of Quinby; thence (4) by land formerly of Bernard Rogan South twenty seven degrees and fifteen minutes west one hundred and two feet to a corner in the public road; thence (5) South thirty eight degrees and forty three minutes West ninety one feet to the aforesaid easterly right of way line: thence (6) along said right of way line North Fifty five degrees and fifty minutes West, three hundred and fifty nine and fifty four hundredths feet to BEGINNING. Containing five acres more or less.

Being Lot No. 3 in accordance with a partial survey made by Walter E. Roberts C.E. May 1929. . . .

The line of asterisks prior to the word "BEGINNING" is exactly as it appears in the deed. There are no words missing from the body of the description presented.

This description starts by labeling our parcel as "Lot No. 3" and referring to the southeasterly corner of some "Lot No. 2," but we do not know where these lot numbers came from. Were they derived from a filed map of subdivision, an unfiled plan of subdivision, or a tax assessor's records? Without this information, we don't know where "Lot No. 2" is in order to begin at its southeasterly corner.

The first course crosses "the public road." What is the name of this road? How wide is it? At what point in the first course do we intersect the road? After crossing the road, we pass "on the southerly side of a blazed Ash tree on the easterly side of said road." This physical call is useful, as evidence of blaze marks often last a significant number of years, and we also know what kind of a tree we are looking for. The "more or less" distance to the terminus of our course makes life difficult, as there is neither an adjoining tract nor a physical marker called for to know where to end this line.

The second course changes measurement units from feet to chains and links. This is a clue that we are working with some kind of composite description, one that was not based on a survey of the entire tract. In fact, the closing paragraph of the deed tells us that this is indeed a "partial survey." This means that only the newly created lines were surveyed, and not the perimeter of the entire parent tract.

The "partial survey" label also explains why Course 1 contains a plus or minus distance; it is a new line which runs from the corner of newly created Lot 2 to the perimeter of the parent tract, whatever that distance may be. A new line created long after original surveying techniques and instrumentation are no longer employed will be reported in new units of measurement. Since Course 2 is in the original chains and links, we know that it is an unchanged course from the original undivided parent tract. The same analysis can be applied to Course 3, which is also reported in chains and links.

We find physical and record calls at the ends of both Courses 2 and 3: a tree for the former and a "stone heap on top of hill in line of land now or formerly of Quinby" for the latter. This combination of physical and record references turns out to be a valuable clue to tying this deed to the ground and to adjoining deeds. In fact, this "hill" is a ridge rising about 70 feet above the nearby roadway, and this particular stone heap was found in the field almost immediately, but not until some additional deed research pointed us in the right direction.

Course 4 reverts to feet, so we know we are once again running along a new line of subdivision. Somehow we end at "a corner in the public road," which gives rise to more questions. Which public road are we in? By "in the public road," is it the centerline of the right-of-way being referenced? Perhaps this is meant to be in the centerline of the traveled way? Or are we just "somewhere" in the road, not necessarily at the center of anything? (In this case, the line did, in fact, run diagonally across the road.) Statutes or local custom sometimes may assist in the interpretation of deed verbiage, but such help was not available in this instance. The term *in* merely means "surrounded by" or "within," leaving our questions to be answered by fieldwork and additional deed research.

The fifth course does specifically run to the right-of-way line of the road, and although we still do not know the width of that right-of-way, by the deed's phrasing we now know definitively that the blazed Ash in the first course is on the legal right-of-way line and not just on the side of the physically traveled way. Our fifth course is another new line, in feet and hundredths of a foot, and Course 6 runs along the road right-of-way line to return us to the tract's beginning point.

Still uncertain about the location of our tract or the history of some of the referenced markers in what we have read so far, we dig further and come up with an older description.

Deed 3 (dated 1815)

... situate in the Township of Amwell, bounded and described as follows: Beginning at a hickory in a line of Nathaniel Larson, corner to Barton Phillips and thence along said Phillips line (1) North thirty six degrees west, eight chains to a heap of Stones in Samuel Well's Line, corner to John White's Lot, thence along the Same (2) North fifty four degrees east, seven chains and Ninety One Links to a heap of Stones, corner to said Lot in a line of said Rodman's other Lands thence along the Same (3) South thirty five degrees east four chains and

thirty six links to a heap of Stones in James Quinby's line thence (4) South
twenty eight Degrees and thirty minutes west eight chains and Sixty five links
to the point of beginning, containing four acres and eight tenths of an acre
of Land.

Due to the later evolution of new municipalities and township boundaries, the location of this deed in Amwell Township does not present a
discrepancy from the 1929 deed. Delaware Township was divided out
of Amwell in 1838, and while Kingwood Township was first formed
in 1749, its boundaries continued to shift through 1876.

The description begins with three references: ties to two adjoiners
(whose deeds should now be searched) and a hickory. Unfortunately,
the tree size is not given, and the described tract is in a semiwooded area.

Course 1 references its location to two adjoining tracts, and is terminated by a physical monument. A "heap of stones" may not sound
impressive, but in fact all of the heaps called for in this description are
neatly piled circles, about six feet in diameter, of large stones clearly
placed by human effort and not the result of geophysical phenomena.
The heaps still existed in 1990 when the parcel was resurveyed. Courses
2 and 3 also refer to adjoining tracts that can be found in the public
record, and both courses terminate at heaps of stones.

Additionally, we find a record link between this tract and the one in
the previously discussed deed, so that we can begin to plot our deed
mosaic: the "heap of Stones in James Quinby's line" in Course 3 ties
back to "a stone heap on top of hill in line of land now or formerly of
Quinby" in the 1929 deed, although that later deed was more definitive
in terms of terrain.

While the geometry of the 1815 deed is mathematically the worst of
the three discussed, misclosing by a greater distance than either of the
two later deeds, the descriptive aspects of this document made it the
most valuable. The clear calls for physical monumentation made those
features readily identifiable in the field. Calls for adjoining owners also
simplified further record research and verified that there were no intervening strips of land between our surveyed tract and adjoining parcels,
no matter what the "metes," or measured bearings and distances, might
lead us to believe. The geometry and mathematical aspects of the 1815
deed are easy to correct and update, but other descriptive features cannot be replaced if lost over the years. Each time we cite found or set
markers, identify distinguishing features of the terrain, and reference
adjoining owners and deeds, we preserve evidence of the lines for future surveyors, and more importantly, for the future landowners whose
property rights we are licensed to protect.

5.5 REFERENCE TO PLATS IN DESCRIPTIONS

Any time that a survey, plan, or other document is referenced in a description, it becomes part of the description and incorporated into it as if physically attached. For this reason, we find numerous descriptions simply referencing a subdivision filed in the county hall of records as the basis of location. In the best of such circumstances, the full title of the plat, any recordation information (plat number, date of filing, perhaps the book or drawer in which it was filed), and the lot number within a specified block are all clearly identified. In this manner we can recover the document on which the description was based to find all the easements, monuments, and other conditions shown on the plat that are incorporated into the description by the mention of the plat.

However, there is a danger in merely recycling such a description. It is always possible that conditions have changed since the original platting, and the original plat or other reference document is rarely, if ever, updated to accommodate such changes. For example, some of the lots could have been further subdivided, either to create new buildable lots or to convey the subdivided area to an adjoining lot to increase the second tract's dimensions, creating a consolidation also not shown on the original document. Some of the easements may have been extinguished or moved, or more recent easements added to the area within the platted subdivision. It is even possible that roads are added or extinguished. For such reasons, proper understanding of true conditions on the grounds for deeds based upon referenced plats or other documents requires more than merely recovering those plats or documents. We'll look at a couple of examples as the significance of this point.

The first case, *Bailey v. Ravalli County* (653 P.2d 139, Supreme Court of Montana, 1982) raised a question about whether or not the road abutting Bailey's property had been closed, and if so, what private rights he had in it. Bailey had prevailed in the District Court in his claim of title to half of the roadway that he said Ravalli County had abandoned. The trial court, which also upheld Bailey's claim, had found that the county had officially closed the road in 1944 and therefore title to the land beneath the road had reverted to the abutting landowners, each receiving to the center of the formerly platted road.

The road in question had been shown on a plat filed and recorded in 1909, with a certificate of dedication on the plat that stated, "The land included in all streets, avenues, alleys, parks, and public squares shown on said plat are hereby granted and dedicated to the use of the public forever." In 1944, a petition to close some of the platted roads resulted in the appointment of road viewers, who recommended the closings, and the county recorded those closings with the following statement: "We find the roads are not being used as public roads and are not likely to be used as such, and closing of said roads would not inconvenience the public in any degree. . . ."[7]

After 1944, the adjoining landowners did not pay taxes on any part of the roadway, although no portion of the roadways was enclosed until 1980, when Bailey erected a fence along the former road's centerline. Then the county commissioners directed him to remove the obstruction to the roadway.

The questions before the court were whether or not the statutory dedication of lands designated for streets and alley in the 1909 plat created a public highway, and if so, if the plat vested title in the county. The statutes in effect at the time of dedication provide the answers. At that time, the most current statutes (from1895) said: "Every . . . grant to the public . . . marked or noted as such on the plat of the city or town, or addition, must be considered to all intents and purposes, as a deed to the said donee."[8]

The Montana Supreme Court thus found that the plat did comply in making its dedication, that the dedication was for a "public highway," and that the plat dedication was equivalent to a right-of-way deed under which the public acquired only the right-of-way and "incidents necessary to enjoy and maintain that public highway."[9] This language signifies that only an easement right had passed to the county, rather than full fee title.

In further explanations, the court confirmed that while the dedication on the plat in 1909 did create a public road or highway, the platted road had been closed and officially vacated by the Ravalli County Commissioners on May 3, 1944, when the board acted on the road viewers' recommendations to close the road, in accordance with the petition (although the court's written opinion used the word *abandoned,* which merely means cessation of use in a manner to connote no further interest). Furthermore, because the original dedication was only a grant of use and not of title, title had never been vested in the public and therefore the dedication only provided an easement over private land for public benefit. Any lawfully vacated or abandoned highway ceases to be a highway, and therefore, as the public merely has an easement of

---

[7]653 P.2d 139 at 141.
[8]653 P.2d 139 at 141.
[9]653 P.2d 139 at 142.

way, the title reverts to the owners of fee unencumbered or "discharged from the servitude."

The significance of this case expands on the question of who owns the soil beneath roads shown on plats. The presence of notations on a plat that state the nature of easements (in this instance, for use only, and for specified purposes) must be supplemented with recitations of local resolutions or other actions that change the status of dedicated rights-of-way. Generally, a buyer of a lot who receives a description that only references a plat has the right to rely on the roads shown on that plat as open for use, as part of the contractual understanding of what is being acquired. But changes subsequent to recordation of a plat must be included in such a simple description in order to provide the full picture.

We should always remember that a plat offers a dedication, but until that dedication is accepted, whether by a signature on the plat by an authorized representative of the accepting party or by a deed, resolution, or other official act, there is no transfer of interest in roads, conservation easements, utility rights-of-way, or any other area shown on the drawing.[10] The plat itself merely describes but does not convey any interests. It is the responsibility of the person writing a description based on a plat to reference the conditions shown on the plat, providing notice of the exceptions and encumbrances shown on the plat, as well as to research and report the status of those exceptions and encumbrances.

Underscoring this last statement and illustrating the kinds of problems created when a description does not contain full disclosure is the case of *MacBean v. St. Paul Title Insurance Corp.* (405 A.2d 405, Superior Court of New Jersey, Appellate Division, 1979). This is a case against the title insurance company, but based on a survey showing a road bordering Catherine MacBean's lot without mention of that road's being unopened, unpaved, and available for future sale to someone else.

The description of the property MacBean bought and had surveyed in 1973 was as follows:

Beginning at a point in the Westerly line of Warren Street, said point being 1300 feet on a course South 11 degrees 35 minutes West from a concrete monument at the Southwesterly corner of said Warren Street and Bay Avenue; thence running as the Magnetic Needle pointed in the year 1950, along the Westerly line of said Warren Street,

---

[10]It should be remembered that an offer for dedication does not necessarily expire even if decades pass without formal acceptance. Furthermore, if the public uses a platted park, street, or other "public area" in accordance with the plat's designation, a common-law dedication can occur through the same legal reasoning as in prescription.

South 11 degrees 35 minutes West 100.00 feet to a point in the Southerly line of the whole tract;

North 77 degrees 54 minutes West 100 feet to a point in said Southerly line of the whole tract;

North 11 degrees 35 minutes East 100.00 feet to a point;

South 77 degrees 48 minutes East 100 feet to the place of BEGINNING.

The following certification appeared on the survey plan:

I certify that this survey has been made under my immediate supervision and that it is correct, as per record description, and that there are no encroachments either way across the property lines, except as shown.

While not evident from the description, the survey depicted what appeared to be a corner lot (see Figure 5.4). Warren Avenue, shown along the easterly line of the tract, was in fact a paved and dedicated road. The survey also showed what the court referred to as a "purported public street" along the northerly line of MacBean's lot, 50 feet wide and labeled "Delaware Avenue." It is this road, which was not mentioned in MacBean's deed, that the suit addresses. Delaware Avenue on the east side of Warren Avenue was paved, but the portion adjoining the MacBean tract was unpaved and uncurbed, conditions that did not appear on the survey. MacBean had her house and garage constructed facing this portion of Delaware Avenue.

In 1975, a letter from the Township Engineer was MacBean's first alert to the fact that Delaware Avenue west of Warren Street was not a public street but a privately owned lot. Shortly afterward, a house was built on that part of so-called "Delaware Avenue" adjoining MacBean's house and lot, facing Warren Street. This meant that the front of MacBean's house now faced the side of the new neighboring house, and that Ms. MacBean could no longer use her driveway.

Failing to locate and "obtain satisfaction" from the surveyor, MacBean turned to the title company. The court noted, "We stress that this case does not require an analysis of whether the surveyor, who is not a party, would be liable to plaintiffs for misrepresenting the existence of an abutting street."[11]

This is not to say that surveyor liability could not be pressed in other cases when the arguments presented at trial are focused just a little differently. MacBean's title insurance policy included standard coverage for "lack of a right to access to and from the land" and lack of

---

[11]405 A.2d 405 at 407.

RE: Catherine MacBean v. St. Paul Title Ins.   A-3018-77, decided 7/11/79

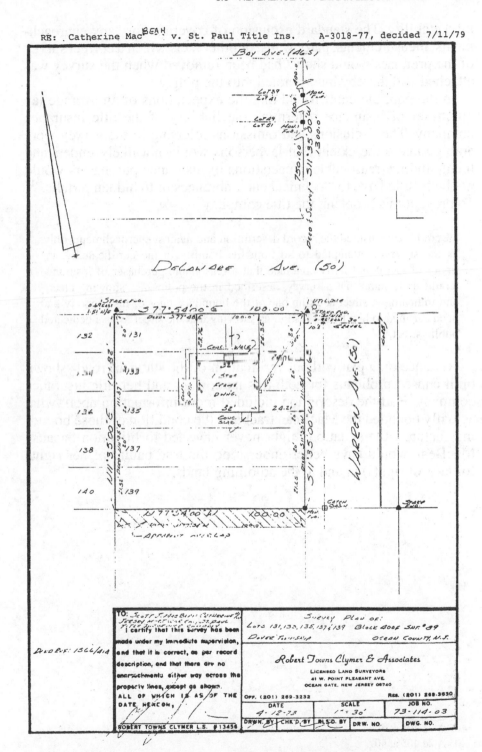

FIGURE 5.4 *MacBean v. St. Paul Title Insurance Corp.*

marketability. The standard exclusion of coverage for "any encroachments, measurements, party walls, or other facts which a correct survey of the premises would show" had been removed when the survey was attached and thereby incorporated into the policy.

In the end, the court found that the expectations of an average lay purchaser of insurance determine the liability of the title insurance company. The exclusions and omissions referring to the survey were ambiguous to the extent that laypersons would not likely understand them, and the reasonable expectations of insurance purchasers would not be fulfilled due to "technical encumbrances or to hidden pitfalls."[12] To prevent this problem, the title company

> ... could have insured the record description and against encroachments only, as the surveyor attempted to so limit his liability in the certificate. ... [A] finder of fact could fairly conclude that a reasonable purchaser of insurance would understand that a survey, described in the policy as "showing clear," and indicating a street abutting one of the boundary lines of his property had been certified to him by the insurer as showing that the street to be a dedicated public street.[13]

Reliance on a plat without verification of the status of roads shown on it caused problems for both this landowner and her title insurance company. Had the description included more information about what actually bounded the MacBean tract and the condition of those bounding features, the situation might never have led to litigation because MacBean would have better understood the true nature of her rights (or lack of rights) in and to the adjoining lands.

---

[12]405 A.2d 405 at 408.
[13]405 A.2d 405 at 409.

CHAPTER 6

COMPOSING, COMPREHENDING DESCRIPTIONS

6.1 GENERAL

Within a deed, a well-written and modern land description consists of four major elements following the general outline of:

1. Caption
2. Body or narrative
3. Qualifications
4. Closing and references

Even if of the "boilerplate" variety, as often occurs with a utility company or a state highway department, a deed will always have a caption, which provides a general location of the site, and a body or narrative describing the property more specifically (although sometimes not significantly more so, depending on the skill and care of the preparer).

Qualifications to a deed description come in a wide variety, each serving to further explain the intent of the parties in the conveyance of real property interests. Qualifications may add to the deeded rights, subtract from them, restrict full exercise of them, or reserve certain rights in favor of the grantor or others.

Items 3 and 4 may either be absent or addressed in only the most cursory manner, depending on mass document preparation (as for right-of-way acquisitions), the parcel's history, and specific site conditions.

Land descriptions that adhere to this outline as much as possible have proven to be clear, concise, and unlikely to suffer misinterpretation at a later date. Each item should be distinctly separated from the others and not allowed to blend or flow together. Whenever possible, each part of the description should be limited to a single paragraph, with the exception of the body, for which each course may begin on a separate line for ease of reading. When multiple deletions, encumbrance clauses, or augmenting clauses exist, each of these may also be a separate paragraph to more fully include descriptive language. When merely referencing source documents rather than including complete descriptions of each one, these deletions, encumbrances and augmenting clauses can be presented in list format, prefaced by a statement of how many items will be in the list. This arrangement avoids the persistent problem of fragmented documentation. Missing pages or portions of a page can be significantly more difficult to detect when document elements are split between multiple paragraphs.

There are times when the ultimate user will dictate the structure of a description. In such instances, it is still the description writer's responsibility to preserve evidence of boundaries and document the intent of the parties involved in the real property interest transaction. The reason for a particular structure may be to keep a large number of documents related to a single project uniform by providing a template, such as may be needed in the acquisition of pipeline easements through numerous private properties. While the dictated format and certain terminology have to be employed, there is no reason to write a less careful narrative.

The goal of any land description is to convey the intention of both parties involved in the transfer of real property interests. The land description should be written under the assumption it will be the sole surviving document identifying that real property parcel. Vital to that intent is the concept of the hierarchy of calls.

6.2   HIERARCHY OF CALLS

In the context of legal descriptions, "calls" are references to evidence of the boundaries of the tract. In other words the evidence is "called out" in the description. This is significant, because evidence found in the field or through record research that is not referenced in the written and recorded description of the property in question cannot be assumed to reflect the intention of the parties involved in a conveyance of real property interests.

A written description in a deed is meant to provide notice to the world of the contract or agreement between the grantor, or seller of property interests, and the grantee, the recipient of those interests. Evidence of the boundaries of those interests may not always be clear, or even present, in the description, and the result can be ambiguity in determining the original intent of the parties. In such instances, the courts have established a relative weight or ranked significance of the various kinds of evidence that may help to clarify that intent referred to as the *hierarchy of calls* or *hierarchy of evidence*. Applying this ranking of calls in a deed, or to any evidence of real property boundaries, is only allowable to overcome ambiguities, but never to introduce a new interpretation or to conflict with other clear statements in a description. Land surveyors apply the ranking in this hierarchy to the data compiled and evidence recovered during a land boundary survey.

The hierarchy of calls has been established through law, regulation, and judicial proceedings, and its roots in English common law means that it is fairly uniform throughout the various United States. The following is the most common ranking of evidence to determine the written intent of the parties.

- Call for a survey upon which the conveyance is based, or reference to a map or deed
- Natural monuments
- Artificial monuments (includes both man-made monuments and record monuments such as calls for an adjoining landowner's boundary)
- Directions and distances
- Area
- Coordinates

Note that the first three items on this list can be classified as "bounds," or limits, while the last three items are "metes," with *mete* being related to an ancient English term meaning "measure." Measurements and calculations are not as direct a means of conveying the intent of the parties as simply pointing to a map and saying "I am selling you this lot" or walking the site and saying "I am buying the land between this tree, that boulder, and the creek over there." Because area and coordinates are derived from measurements, they trail distance and direction in terms of evidential weight.

As surveying methodology changes, so does the hierarchy of calls and evidence in a description. In older texts and cases, often direction has held more weight than distance. This was based on the difference in accuracy of instruments used to measure directions as opposed to those used to measure distances. In more modern times, however, as instrumentation has changed the practice of surveying dramatically from when the use of compass and chain was the norm, the differences between accuracies of measured directions and measured distances have been eliminated. Therefore, when reading older deeds based on older forms of measurement, there may in fact be reason to have greater confidence in reported directions than in reported distances, while a description based on a recent survey is read with these elements having equal significance.

There is also the possibility of local differences in the ranking. For example, in an area with a lot of ore in the ground, local attraction might have made compass readings in older surveys less than consistent, so that measured distances under such circumstances would be considered better evidence in retracing the boundaries of a property. However, there may also be a local difference in how distances are measured and reported, as addressed in Chapter 5.

And so the hierarchy is not cast in stone. The overriding principle in interpreting any description of real property is to follow the *intention of the parties*. Any of the categories of evidence can be challenged for authenticity or veracity. None of the evidence can trump a clear, concise, and unambiguous presentation of the intention of the parties who created the parcel. Land descriptions must incorporate *all* of the known and pertinent items in the hierarchy of calls in order to properly convey the intention of the parties. Any evidence of boundary location discovered or established during the course of performing a land boundary survey that clarifies the intention of the parties should be reported on the plat of survey and documented in the written land description.

6.2.1 Elements of the Boundaries

The hierarchy of calls, so important to the recovery of property corners, must be supported in the narrative of the description. If the description does not include references to evidence of the boundaries, the result is that subsequent readers will not find it as clear as the original signers of the deed meant it to be. Such ambiguities may end up settled in court on the basis of equity or legal principles when the original intent is not evident in the written record, yielding a very different outcome than what was meant by the writers of the description. Worse, juries and/or judges who lack any understanding of real property interests often settle ambiguities erroneously. The best protection of real property rights is a clear description that includes all the evidence available at the time of writing. This preserves the intent for the future, when the original parties and even some of the original evidence are no longer available.

Each course of a description follows a single line of the boundary, whether that line is straight or curved. Beyond the direction and length of the line, each course must expressly include any physical evidence and any record evidence of its location. This preserves the validity of later interpretations based upon the hierarchy of calls. Evidence that is not called for in a description often will not be considered by the court.

Let's look at a few examples from our hierarchy of calls.

6.2.1.1 Call for a Survey on Which the Conveyance Is Based, or Reference to a Map or Deed

Did the parties intend a line to run along a common boundary with another tract, or to create a new line as a subdivision of a larger parcel? Supply the name and date of any recorded survey or subdivision plat on which the parties relied in defining the conveyed tract. If the land is the same tract as was transferred by a specific deed, reciting the book and page in which the deed was recorded provides similar evidence.

Once referenced, a document is considered part of the deed, as if attached to it. In jurisdictions where recording surveys is neither required nor standard practice, copies of surveys are often available from the preparer or a later repository of those records as a matter of professional courtesy or for the cost of retrieval from the file and reproduction. Provide the date, title, preparer's name and company, and any other identifying information that will help in locating or recovering it.

Sometimes a boundary has been adjudicated, in which case a court decision may need to be referenced if the decreed final boundary has not been recorded with the county's deeds. In such an instance, the decision has the same weight as a survey, map, or deed, and is to be cited by its docket number, date, and any other record information that will help others to find the original document.

In some localities, land can legally be transferred by mere reference to a tax map parcel identification number or its tax block and tax lot identification. However, a tax map is an assessment tool, not a survey. Even if the tracts shown on a tax assessment map are based on deeds and surveys, such a map cannot be relied upon with the same confidence as a signed and sealed survey prepared by a licensed professional surveyor. A tax map provides, at best, only secondhand information and interpretation of the original surveys and deeds from which it was prepared. Tax maps do not carry the same authority as deeds and surveys, and a tax map's mere depiction of supposed conditions does not perfect dedication or vacation of public or private rights.

Even in places where laws mandate that only a licensed professional surveyor shall create the base mapping for or perform updates of a tax map (the tax map itself being based on surveys by others), this assessment tool is not as current or as complete as a full survey. Conveyances occur more frequently than tax map updates, and descriptions containing only references to a tax map parcel have created terrible problems for the grantees of such tracts.

While we try to avoid absolutes, in this case it is safe to say, "*Never rely solely on a reference to a tax map in describing real property.*" If adding a clause at the end of the description to say that the tract is also known as Tax Lot 93 in Tax Block 47 of the Borough of Hometown in Upper County, include the date of the tax map referenced (accompanied by the date of last revision) to establish conditions at the time the description was written.

6.2.1.2 Call for Natural Monuments

Unless specifically stated otherwise, reference to a natural monument such as a tree or a ditch is generally understood to be at the center of that natural monument, just as a reference to a capped steel pin is understood to mean the center of that pin. Any other intent must be clearly and specifically stated.

Of course, there are some well-understood and well-adjudicated exceptions to this generalization. A call to a body of water does not always convey interest to the center of that water. Tidal bodies of water are just one example of such an exception. Depending on the jurisdiction, the land beneath tidal and navigable bodies of water belongs to the sovereign body (state or federal) in trust for the citizens. The reason is based in English law, when the highways and byways between two towns were sometimes waterways rather than terrestrial roads. The crown held the title in these traveled ways and protected the rights of the public to use them for navigation and commerce. The federal government and various states on this side of the Atlantic Ocean have adopted a similar approach, although ownership by the sovereign may be either to the high water mark or the low water mark of the water. If to the low water mark, the upland owner is generally subject to the rights of the public to access the water in the area between the low and high water marks.

In the case of sectionalized public lands, distances and bearings to and along navigable rivers, streams, or the shores of lakes or the sea were measured in reference to a *meander* line. Real property rights did not terminate at these meanders, which were property lines only when so named in deeds. The property rights extended past the meander line to the water body according to the dictates of riparian boundaries that are defined by the nature of the bounding water body.

There are other possible intended boundaries for land bordering water. A reference to the *thread* of a body of water indicates the intent to run to the middle of the stream or river, giving each abutting upland owner equal access to the water. The thread of a river or stream is defined as a line equidistant from the water's edge on the two sides of the river or stream during the water's ordinary height or elevation. Sometimes the thread of a river may be defined as equidistant from the ordinary low water mark on each bank. In either instance, state statutes should be consulted to establish the accepted definition within a given jurisdiction. Furthermore, every state has either statutes or constitutional articles identifying its boundaries, and these should be researched when surveying along state lines to assure consistency in the defined common line, just as one would do when researching deeds for privately held lands.

When the bottom of the stream or river is not of uniform depth, sometimes mere splitting along a centerline will result in one abutting owner's not being able to navigate in a boat or ship because of the shallows. In such instances, the division line maybe called to the *thalweg*, or the center of the deepest or chief navigable channel in the waterway. This is a common boundary between states separated by a river that both wish to use, so that each state has the right to the possibilities of commerce and navigation offered by that river.

There can also be local differences in the language of older deeds, and here is where familiarity with regional custom and practice is important in understanding the intent of the original writers of the description. In the case of *Padilla v. City of Santa Fe* (753 P.2d 353, Court of Appeals of New Mexico, 1988), a deed originally written in Spanish conveyed land *"a las lomas,"* meaning "to the hills," and the parties in this suit argued over the definition of where that call actually ended. Did it mean "to a point on the top of the hills," thereby including all the land on the side of the hill? Or did it merely run to where the hills began, being "to a point at the foot of the hills"?

The answer lay in a local interpretation of the phrase, which the appellate court ruled should be incorporated into the interpretation of this call for a natural monument. Local custom was to ensure that grantees would acquire land "far enough up the slope of a hill to allow houses, barns, etc. to be built without wasting any valuable arable land."[1] Furthermore, testimony revealed that local custom was to call "to the foothills" if the writer of the deed intended to transfer only to the bottom of the foothills. The New Mexico Court of Appeals overturned the trial court's preferred reliance upon the deed's cited distances, and instead found the intended boundary to be at the crest of the first hill above the road, rather than at the "toe of the hill" as the city claimed.

6.2.1.3 *Artificial Monuments*   Artificial monuments include both man-made objects and record monuments, such as calls for an adjoining landowner's boundary.

Were the iron rods mentioned at the tract's corners part of the original survey creating the land being conveyed, or were they found and assumed to mark the corners of an adjoining tract? When writing a description, any physical evidence should be identified as either found or set. This assists anyone tracing the title of land to its origins in determining whether what was found in the field was part of the original creation of the tract, a later revision to its boundaries, part of an adjoining survey, or a result of a recent survey of the property in question.

Beyond markers at the actual corners of properties, there may be reference markers or accessories to the corners that help locate corners in the field. A corner may be inaccessible or impossible to mark for any number of reasons. Perhaps it falls in the center of a river. Perhaps it is in a quicksand area. Maybe it is within a stone fence row. In such instances, the surveyor often sets an offset marker on one of the lines leading into or away from the corner as a witness to that corner. Sometimes the surveyor may note the nearby presence of an existing feature of a permanent or semipermanent nature, such as a tree or the corner of a building. In such a case, a tie bearing and distance from the actual property corner to the clearly described accessory will help pin down the location of that true corner.

A reference to an adjoining landowner's boundary is considered an artificial monument because such a line exists in written records as a human creation rather than as a naturally occurring feature. Such reference can establish a junior/senior relationship between the tracts, with the junior or later-created tract referring to the line of the senior or first-existing tract; the citations should never be the other way around. If the boundary is intended to be the common line with Charles Goodson as that line was confirmed by Mr. Goodson's deed, include the recorded book and page for his deed as well. Such practice will help future deed

---

[1]753 P.2d 333 at 357.

readers and surveyors find relevant written records during the process of tracing title back to its roots.

Properties that front along public roads deserve special attention. The title in the road itself may belong to the owner of the abutting property, subject to the rights of the public to cross over it for road purposes, or it may belong to a government entity, held in trust for the public's use. Generally, unless specifically excluded from the deed, a tract that runs "to" the road includes title to the center of that road, just as the call "to a marble monument" runs to the center of that stone. However, if title does not belong to the abutting owner, then running "to" the road can legally only transfer title to the right-of-way line of the road.

Here is an instance in which careful thought about future readers of the description helps us to write a more complete and useful document. The width of a road right-of-way may change over time, either widening or narrowing, sometimes not equally on each side of its original centerline. The centerline of a road created as a two-rod road in 1813 is one rod, or 16.5 feet, from each of the two right-of-way lines defining that road. Over time, the road may be widened on one side and not the other, but the original centerline remains. Now the road may encompass 25 feet on one side and 16.5 on the other side of the centerline. Another change to the road may occur on one or the other side of the centerline, but citing the current distance from the right-of-way line to that original centerline as of a certain date (obviously the result of having conducted some thorough research to establish where that lies) is the best way to recreate conditions on the ground and to assist future surveyors in their own research.

Why is this important? Consider the situation in which a city establishes its infrastructure and maps it in relation to the centerline of its streets, or in relation to curb lines in relation to the centerline of its streets. This is not a hypothetical case. The combined sewers of the city of Philadelphia, Pennsylvania, were mapped in just such a manner, and when repairs were necessary a century later, research found that the sewers had been drawn on record maps in relationship to the curb lines that lined the traveled way of the streets.

Over the course of decades, the streets had been widened, not always uniformly on both sides of the original centerline, so that determining the present location of old buried pipes required piecing together a puzzle that had evolved over time, The chronology of street widening, tracking which side of the street had been widened to what extent in which year, had to be compared to the dates of sewer installation that had also expanded the city's sewer system over time.

In some instances, the original streets no longer existed, having been obliterated by parks and a boulevard cutting diagonally through the original gridded street pattern. In order to excavate in the right place to unearth old manholes and pipes, the research and mapping exercise had to be executed carefully. Definitely, the dates and recited right-of-way widths from right-of-way line to centerline played a crucial role in this project.

When describing properties along public roads, case after case has confirmed that title runs to the centerline unless the intent to limit title to the sideline is stated specifically. Further, merely stating that the line of the property runs along the right-of-way is not sufficient to exclude title to the centerline. The case of *Salter v. Jonas* (39 N.J.L. 469, Court of Errors and Appeals of New Jersey, 1877) is cited in cases around the country in support of this point.

The Salters had mapped and divided their land to create lots and streets. This suit debated the ownership of half of what had been a public street in front of a particular lot on that map that had eventually come into the ownership and possession of Jonas. The deed by which the Jonases took title read as follows:

> All that certain lot or parcel of land, situate, lying and being in the township of Bergen, in the County of Hudson and State of NJ, butted and bounded as follows: Beginning at a stake standing at the junction of the easterly line of Rowland Street with the northerly line of Johnson Street, as laid down on the map of said Salter's premises, and running thence: (1) along the northerly line of Johnson Street South 23 degrees 40 minutes East, 50 feet to a stake; thence (2) North 66 degrees east, 100 feet to a stake; thence (3) North, 23 degrees and 40 minutes West, 50 feet to a stake in the said easterly line of Rowland Street: thence (4) along the same South 66 degrees West 100 feet to the beginning.

Rowland Street eventually "became useless" (as the court described it, in this context likely meaning abandoned or vacated), and therefore Jonas had fenced in his property to include the land to the middle of the street. The Salters, as original owners and original grantors, claimed title to the land between the sideline and centerline of the "useless" Rowland Street, based on language of the deed that ran along the lines of the cited roads. The Salters lost both in trial court and in their appeal.

The higher court noted that the street is an appurtenance to the abutting property, something of great value to the contiguous lots but of little value as a separate strip of land. In upholding the trial court finding in favor of the Jonases owning to the center of the streets that abutting his property, the upper court quoted the trial court judge in his observation of Salter's complaint that the Jonas "titles do not extend to the middle of the street, because the lots, as described, are *bounded by the sides of the streets*. But the established inference of law is, that a conveyance of land, bounded on a public highway, carries with it the fee to the centre of the road, as part and parcel of the grant."[2] Nothing short of express words of exclusion will prevent title to the center of a public street from passing to the grantee of a tract fronting on that street.

Ownership of the beds of private streets is an entirely different matter. Such roads are often in the form of easements allowing others to use that means of access, ingress, and egress only with permission of the owners of the soil beneath such roads.

6.2.1.4 *Directions and Distances* Each course or line segment in a description must have a clearly defined direction and distance. For courses that run along wandering features such as a ditch, a general direction of the actual ditch should be noted, and a calculated tie line between the origin and terminus of the course will help future surveyors locate the corners of the tract. For example:

> Running in a general northerly direction along the center of a ditch for approximately 1200 feet to an angle point, said course having a computed tie between its origin and terminus of North twelve degrees seventy two minutes ten seconds East, 1981.37 feet, thence . . .

In the instance of a curved course, enough information must be provided to recreate the course mathematically, to know whether it is meant to curve to the right or left, to establish whether it is tangential to an immediately prior or subsequent straight line segment or if it shares a radial line with an immediately prior or subsequent curved course. More will be said about curved courses further on in this chapter.

6.2.1.5 *Area* While there may be an intent to convey a set number of acres, hectares, square rods, or other area of land, the boundaries of that area must still be established by the parties to the transaction so that the tract is not floating unmoored to the surface of the earth. Even when the stated intent is to convey the east 20 acres of an existing parcel, the limit of those so-called 20 acres is generally a known entity. Therefore, in the instance of a conflict between known boundaries and stated area, the known boundaries will prevail.

---

[2] 39 N.J.L. 469 at 474.

In other instances when a complete metes and bounds description is followed by a statement of area, that area is a calculated product derived from the primary description of the tract's boundaries. As a result, area falls very low on the hierarchy of calls when it comes to weighing boundary evidence.

6.2.1.6 Coordinates

Coordinates are not naturally occurring, having no physical or tangible existence. These numerical values are the calculated results of measurements (which are themselves secondary to referenced monuments), and therefore can at best be regarded as only tertiary evidence of a boundary. While an increasing number of municipalities, counties, and other land use regulating agencies are requiring coordinate values to be reported on subdivision plans and surveys, the recording of documents on which they appear does not alter coordinates' tertiary nature. Public recordation may make accessing the land a faster process, but there are inherent difficulties with reliance on coordinate values.

While it is common knowledge among surveyors (but not necessarily the general public) that measurements of a line may vary over time due to changes in technology or due to varying practices between surveyors, variations in the computation of coordinates in different reference bases is not so well understood. It may be possible that one surveyor utilized a global positioning system (GPS) to obtain coordinate values, while another performed a ground survey from a known United States Coast and Geodetic monument and yet another merely scaled values from a United States Geological Survey (USGS) topographic map or even a private geographic information system. Comparison of values derived from such different sources is a complicated matter, well beyond the layperson and sometimes beyond the abilities of land surveyors who are used to surveying on a relatively small flat surface and not well versed in geodesy and spheroid geometry. Consequently, quality of coordinates may vary widely.

While governments and other agencies may require a surveyor to report coordinates for the corners of surveyed tract, these numerical values hold little or no weight in determining the boundaries intended by the parties to a conveyance of real property interests.

6.3 CAPTION

The caption of a description provides a generalized location of the site, a frame of reference to begin understanding the position of the property.

In the colonial system, the state, county, and municipality are the most common elements included. In the United States Public Lands System (USPLS), this same information is recited, with additional information relating to the specific government survey identifying features such as reference meridian, township, and range. For sites above or below the surface of the earth, it may be necessary to include elevation data in the caption to provide the proper frame of reference.

6.3.1 Land Record System

The caption or initial paragraph is a product of the land record system used to create the parcel. If the parcel is the result of a platted subdivision, then information contained in the caption will be the lot, block, subdivision, and so on, that identifies the parcel. A metes and bounds caption cannot be used for a USPLS aliquot division of land. Section 23 of a USPLS township has been divided into thirds in Figure 6.1, but this nonaliquot division must be described either by reference to a platted subdivision or by metes and bounds.

A colonial caption of the briefest form can be:

All that parcel of land situated in Narberth Borough, Montgomery County, and Commonwealth of Pennsylvania, being more particularly described as follows:

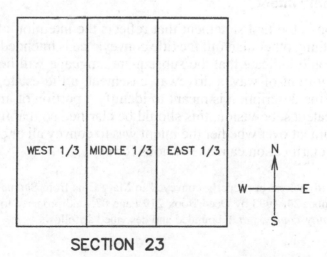

SECTION 23

FIGURE 6.1 Note that the parcels in this example were not developed according to the instructions for aliquot division. If the example plat had developed and recorded so that all three lots were created at the same time, it is a platted subdivision. If the parcels were created at different times, the land record system would be a metes and bounds.

In this example, the general location is established, reserving the specific description for the body of the description. A few other colonial examples follow, sometimes revealing the basis for the details that will appear in the description's body, and sometimes providing only additional general locative information.

All that certain tract or parcel of land located at 200 Kendrick Avenue in the Township of Gloucester, County of Camden, State of New Jersey bounded and described as follows:

All that certain parcel of ground situate in the City and County of Philadelphia, Commonwealth of Pennsylvania, bounded and described according to a survey made by Franklin & Lindsey, Inc. dated November 1928, as follows, to wit:

Description of a certain parcel of land being situated in the Borough of Kinnelon, County of Morris, in the State of New Jersey and being further described as Tax Lot 18 in Block Number 58 on the official tax map of Borough of Kinnelon, Morris County New Jersey, being more precisely described as follows:

All that certain messuage[3] and lot or piece of land being Lot No. 18 on "Plan of Lots of Isaac Warner Arthur", said plan on record at Norristown in Deed Book 278 facing page 38, and situate in the Township of Lower Merion in the County of Montgomery, Commonwealth of Pennsylvania, bounded and described as follows to wit:

6.3.2   Clarify Intent

The caption is the first statement that reflects the intention of the parties. If anything other than full fee title conveyance is intended, then the caption should indicate that the subsequent language will be describing a utility right-of-way, a driveway easement, a life estate, or other interest. If the description is meant to identify a portion of an existing tract to create a subdivision, this should be clarified so that there is no future argument over whether the intent was to convey all or part of the land. This clarification can take various forms:

A one-third interest in property conveyed to Mary Ellis from Samuel Grossi on December 24, 1998 by Deed Book 219 Page 87, said property located in (*municipality, county, state*), bounded and described as follows:

---

[3]*Messuage* means a dwelling and the small area of land immediately surrounding it, not necessarily enclosed. The term is often applied to include the "curtilage," or other buildings immediately surrounding and related to the dwelling, used for domestic purposes.

A 16.5′ wide right-of-way for public utility purposes along the northern boundary of premises located in *(municipality, county, state)* and being more particularly described as follows:

In instances where a recorded plat and the legal description are in complete agreement regarding the boundaries, the caption alone may be sufficient to identify and locate the property without referring to or requiring the body of the description. "Unit No. 119, Common Interest Community No. 39, The Pointe Village Homes, Wright County, Minnesota" is a legally sufficient description. But adding a little more information makes that single sentence even clearer: "Lot 15 on the Revised Plat of Knob Hill Estates, recorded in the Sussex County Clerk's Office in File 21-540 on March 3, 2010," supplemented by the municipality and state where the site is located, is a more complete description.

When any other information beyond the plat reference is to be added, such as identification of markers found and set or information regarding adjoining deeds and owners, such a bare bones description is not adequate, particularly in states where the PLSS and aliquot division descriptions are not employed, and the caption should be supplemented with a full bodied description.

A caption that begins "All of that parcel..." should not have an accounting for a deletion or addition later in the description. If a deletion or addition is present then the caption for a Public Land Survey System (PLSS) tract should begin "That portion of *(parcel)* less than and except *(deletion)*" in the case of a deletion or "All of *(parcel)* and *(addition)*..." in the case of an addition. For a metes and bounds tract, the caption for a tract with a deletion will simply begin, "That portion of *(parcel)* described as follows...."

The caption of a description should be applicable to one specific parcel and no other to meet the definition of "legal description." It should be *sufficient to identify the subject property as a unique entity* without the body of the description. In the case of a PLSS tract, this can be reference to an aliquot division. For a metes and bounds description that is not part of a recorded subdivision, it may require a little more thought and creativity. This can take the form of "being a conservation easement over the land described in Deed Book 135 at page 98," "lands remaining to the grantor after subdivision and conveyance as described in Deed Book 89 at page 42," or "a tract fronting on Pine Street identified as Tax Map Block 8102, Lot 78 on the tax maps of Plainsboro Township."

6.4   BODY

The body or narrative portion of the land description is traditionally based on a survey of the parcel (although in the past the body has sometimes been based on a protraction, meaning a never-surveyed subdivision plat). The survey may have been professionally performed or the result of actions taken by amateurs. It is a commentary of a trip around the boundaries of the parcel.

The narrative of a land description is a documented guide that places a hand on the shoulder of the reader and directs that reader to *location of the real property corners* while following the *boundaries of the parcel.* The narrative must identify the boundaries and describe the corners, including monumentation.

In some instances, the body will consist only of a reference to a public record that more specifically describes the boundaries. For example:

> Being all of Lot 1, Block 3 as shown on a map entitled "Cynwyd Estates," filed in the office of the Montgomery County Clerk on December 10, 1898, as Map Number 345.

No matter whether the narrative is a single sentence and paragraph in grammatical terms or is made up of a longer series of phrases, it is divided into many segments using specialized punctuation and phraseology. These *terms of art* and *key phrases* are utilized to assist the reader in visualizing the configuration and limits of the parcel. The stylized presentation recommended here is the result of centuries of resolving boundary disputes arising from confused, misinterpreted, and fragmented deed documentation.

6.4.1   Point of Commencement

A commencement point provides a frame of reference and orientation for a tract by tying the description's point of beginning to a well-known and accessible object, location, or corner. It should be public in nature, easily found, widely recognized, durable, and recoverable. The commencement point is intended to assist the reader in locating and recovering the property parcel; *it does not define the location of the parcel.* The description of the point of commencement narrows the search for the parcel corner that the writer has chosen to begin the trip around the parcel.

The key phrase "Commencing at" should be immediately followed by the corner or object that is the commencement point. In certain cases the commencement point and the point of beginning can be the same object. Odd or colloquial phrases such as "remote point of beginning" or "first point of beginning," which then lead to "the true point of beginning" are confusing and should be avoided.

Examples of commencement point descriptions:

Commencing at the $^1/_2$-inch-diameter iron pipe found at the intersection of the southernmost right-of-way line of Front Street and the westernmost right-of-way line of Adam Road; thence . . .

Commencing at the brass disk set at the northeast corner of Section 23, Township 12 South, Range 2 West; thence . . .

Notice how the physical characteristics and other information concerning the commencement point are reported. The purpose of such detail is to assist future examiners of the record to actually find the object used in the description. Boundary and right-of-way lines are subject to interpretation or misidentification. If, at a later date, the intersection of street rights-of-way or a section corner referenced as a commencement point were found to actually be in a different place, the recovery of the point of beginning would not be hampered by a incorrect labeling of the of the commencement point.

References to deed tract lines, section lines, and subdivision tract lines may also serve as commencement point intersections.

Commencing at the Southeast corner of lands of Roger Serrano as recorded in Deed Book 296 at page 411 in the Mercer County Clerk's office; thence running along the southerly line of Serrano's land . . . to the point of beginning; thence

Commencing at a mound at the Southwest corner of the Southeast quarter of Section 17, T4N, R1E; thence . . .

A STRIP OF LAND IN LOT 3, BLOCK 3 OF MALONE SUNNY SLOPE SUBDIVISION IN THE CITY OF JACKSON IN CAPE GIRARDEAU COUNTY, MISSOURI DESCRIBED AS FOLLOWS: Commence at the Southwest corner of Lot 2, Block 3 of Malone Sunny Slope Subdivision (5/8″ iron rod); thence N 05°30′00″ W along the East line of Farmington Road, 128.59 feet for the point of Beginning: Thence . . .[4]

---

[4]*Dohogne v. Counts*, 307 S.W.3d 660 at 668, Court of Appeals of Missouri, Eastern District, Division Two, 2010.

6.4.2   Point of Beginning

The point of beginning (POB) must be a location on the boundary of the parcel being described. It is almost always a corner of that property, and it is chosen at the discretion of the writer. The POB established should be the most accessible of the parcel corners. It should be readily identifiable and, if possible, the corner in nearest proximity to the commencement point. This is the boundary corner that will form the starting point for the guided trip around the parcel.

The first occurrence of the key phrase "point of beginning" in a description should include all the data pertinent to that corner. Report the type of object used to monument the corner, whether the monument was found or set, any coordinates or accessories to the corner, and other descriptive data that will help the next surveyor recognize the monument. Such data will also preserve evidence of the corner shared with adjoining landowners. Include a full citation of deed references and ownership if using a record monument rather than a physical one.

Examples of POB phrases are:

> ... to an iron post found on the easternmost boundary line of Lot A and the Point of Beginning; thence ...
>
> ... to a $^1/_2$-inch pipe set on the southernmost right-of-way line of Phillips Highway and the Point of Beginning; thence ...

The second occurrence of the phrase "point of beginning" is the termination of the imaginary trip around the parcel. None of the identifying particulars of the POB is repeated. All information concerning boundaries, adjoiners, corner features, or other data pertinent to the parcel being described should be placed between the first occurrence of "point of beginning" (*Beginning at a concrete monument found ...*) and the last occurrence of "point of beginning" (*... to the point and place of beginning*).

The distinction between a point of commencement and a point of beginning must be very clear, because confusion between them can create wildly disparate understandings about where the property within a description is located. One of the authors of this book had a client disputing his neighbor over the location of the line between them. While in fact there was a discrepancy between their deeds, it was a matter of only a few feet rather than the 1,800 feet that this client claimed. His arguments were based on his knowledge that his property's deed description, which had not changed appreciably in nearly 200 years due to the agricultural nature of the area, began at an oak tree.

However, the deed bore a commencing course leading to the POB for his deed, and the reference for that point of commencement was also an oak tree. The difference between them was that the point of beginning was a red oak at an angle in the ancient road right-of-way line, while the point of commencement was a yellow oak marking the intersection of the common line of two tracts (whose owners were named in the original deed) with the same road right-of-way line. Believing that the point of commencement marked his point of beginning, the client claimed that he had title to over 60 acres instead of only 13, and that his neighbor's chain of title was fraudulent. Needless to say, he was not happy to be enlightened as to the true intent of the commencement course.

Such confusion between the terms *beginning* and *commencing* has been the subject of published litigation for as long as land has been described using these terms. The following description is from 21st-century litigation, based on a description written in the middle of the 20th century. From the units of measurement, it is obvious that the source of the description is older than the referenced survey.

A certain tract of land in the Parish of Livingston, State of Louisiana, in HR 46 T 7 SR 6E, particularly designated and described as *commencing* at the Northeast corner of the 7.50 acre Wm. Winder estate tract of land in the Southeast portion of HR 46, and from said Northeast corner measure North 15.51 chs. to the South boundary line of a 100 acre tract; thence West 6.45 chs.; thence South 15.51 chs.; thence East 6.45 chs. *to point of beginning;* as per survey made by C. M. Moore, Surveyor, on May 18th, 1946.[5] *[Emphasis added]*

While there is nothing wrong with the point of commencement and the point of beginning being one and the same, interchanging the terms as in this example can be confusing as the reader searches for where the tract's boundaries actually begin.

If using both terms, the unity of the point must be unmistakable:

Commencing at the Southwest corner of the Northwest quarter of the Northeast quarter of Section 34, T6S, R6E, which is also the point of beginning:

To circumvent the confusion between commencement and beginning points, description writers can create a visual difference on the printed page as well as a verbal distinction between these two locations. Possibly, we can even omit specific references to the commencement point. The following example uses different formatting to differentiate

---

[5]*Winder v. George,* 973 So. 2d 180, Court of Appeals of Louisiana, First Circuit, 2007. Unpublished, online from Lexis.

between courses running from the point of commencement to the point of beginning and courses that describe the tract in question. Commencing courses are inset rather than being left justified. In this example, they are also labeled alphabetically rather than numerically to further differentiate them from the rest of the description. While the point of commencement is never named as such, it is readily identifiable as distinct from the point of beginning.

> **BEGINNING** at a point marked by a capped iron pin found in the southeasterly line of Lot 34, Block 1001, lands now or formerly of The County of Somerset, being the following course from intersection of the northwesterly Centerline of Belle Mead – Blawenberg Road, also known as County Route 601 (45′ wide right-of-way) with the southeasterly line of Lot 35, Block 1001, lands now or formerly of Richard M. Grosso, Jr.:
>
> (a) Along the said southeasterly line of said Lot 34 and Lot 35, Block 1001, S 83° 19′ 07″ W, a distance of 2122.38 feet
>
> and from the Point of Beginning running; thence
>
> 1) Along a new line through Lot 42, S 02° 56′ 35″ E, a distance of 585.96 feet to a point marked by a capped iron pin set . . .

6.4.3 Elements of the Boundaries

In section 6.2.1 we discussed the elements of the boundaries as they relate to the hierarchy of calls, establishing their weight as evidence in general terms. In this section we look at how to preserve that evidence in the body of descriptions we write.

6.4.3.1 Monuments Surveyors should always remember that a call for a survey in a description is a call for any monuments set on that survey. This guides how we read descriptions citing earlier surveys and the care with which we should include references to surveys in our own written descriptions. When a description references a record line or a physical monument, these become controlling calls that must be identified and located. The rest of the tract must be fitted to these calls.

For any monument to control the location of a boundary line, it must be referenced in the description. If it is found in the field but not mentioned in the deed, the physical monument has no "pedigree," no history or basis on which it can claim authenticity regarding the intent of the parties to the conveyance. As a corollary to this statement, when monumentation is referenced in a description, it must be identified as either found or set, thereby giving some frame of reference within a tract's chain of title. Including the features of physical monuments found or newly set assists later surveyors determine if what they find in the field is original as called for in the written record, identifiable (having well-defined characteristics), and undisturbed (in the reported location) in order to rely on it as evidence of a corner location.

6.4.3.2   Adjoining Owners   A description may reference a common line with another owner, which requires the reader of the description to do a little extra research to determine where that contiguous owner's property is located and what physical markers monument the common line. However, a little care should be exercised when including calls to adjoining lands.

By expressly calling the terminus of a course to be at that particular line's intersection with another line, the second line becomes a controlling factor in setting the limit of the line. But we should remember that not all lots are created alike. Instead, we have lots created simultaneously, as in a platted subdivision (or at least in most jurisdictions the platting of lots creates them simultaneously), and we have lots created sequentially, as in a series of divisions. In the first instance, all lots have equal rights and none is junior or senior to any other. But In the second, the first parcel carved from the parent tract is considered senior, with later parcels divided from the same parent tract being junior to that first parcel and to each other in the order of their creation. For this reason, a senior tract should not be described as ending at the line of a junior tract, for in fact the law sees the situation in exactly the reverse order: a junior line must end at the line of the senior tract. The senior parcel gets its due, and any excess or deficiency in area is then allocated to the junior parcel.

6.4.3.3   Directions—General Orientation   The directions in a description begin with a *general orientation* followed by specific instructions. The purpose of the general orientation is to assist the reader in interpreting more specific instructions. The general orientation also reduces the incidents of misinterpretation of data or, more important, blunders by the writer. The purpose of the general orientation is to turn the reader in the correct direction and to prepare for the specific direction to follow. This orientation is usually reserved to the cardinal directions of "north," "south," "east," and "west." The addition of the suffix *-ly* to the cardinal direction identifies the word as a general direction and a key word. *Northerly direction* (or *northwardly*) is a general orientation, not a specific bearing. The importance of a general orientation is best demonstrated by example.

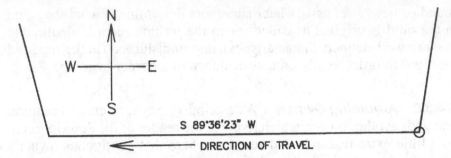

FIGURE 6.2 The bearing of the labeled line is close to a cardinal direction, affecting how we describe the line's general direction.

In Figure 6.2 the writer (and the reader) may be confused as to the correct form of the bearing.[6] The line runs very close to a cardinal direction. Clearly, the general direction of travel is in an easterly direction, but the bearing is reported as running southwest. It is not possible travel southwest in an *easterly* direction. Therefore, the reverse version of the bearing, North 89 degrees 36 minutes 23 seconds East, must be employed. The example line segments are repeated here with the general direction key phrase in italics.

> ...; thence, *in a southerly direction* along the westernmost right-of-way line of Kings Highway, South 15 degrees 32 minutes 15 seconds West, a distance of 325.62 feet to a $^1/_2$-inch-diameter iron rod set on the northernmost boundary line of Lot 23, Kings Estates; thence...

> ...; thence, *in an easterly direction* along the southernmost boundary line of the Smith Tract, North 88 degrees 15 minutes 23 seconds East, a distance of 100.12 feet to an iron pipe found on the westernmost boundary line of the Schultz Tract; thence...

Another general orientation term is the suffix -*most*. This suffix is also used with the cardinal directions to identify one line out of a set of boundary lines. For example, street rights-of-way are most often parallel or concentric lines. A reference to a typical "through" intersection of two streets provides four possible corners.

The simple statement "...at the intersection of Buford Road and King's Highway" does not distinguish between the four possible corners shown in Figure 6.3. Compare this statement with "...at the intersection of the northernmost right-of-way line of Buford Road and the westernmost right-of-way line of King's Highway." The addition

---

[6]See Chapter 3 for more detailed discussion of directions and bearings.

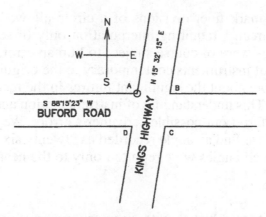

FIGURE 6.3 Road intersections provide multiple corners that must be distinguished in a description.

of general orientation terms to specify which of the public rights-of-way are intended makes it clear that the point labeled "A" is the corner intended.

The example line segments are repeated here with the key phrase in italics.

> ...; thence, in a southerly direction along the westernmost right-of-way line of Kings Highway, South 15 degrees 32 minutes 15 seconds West, a distance of 325.62 feet to a $^1/_2$-inch-diameter iron rod set on the *northernmost boundary line of Lot 23, Kings Estates*; thence...

> ...; thence, in an easterly direction along the southernmost boundary line of the Smith Tract, North 88 degrees 15 minutes 23 seconds East, a distance of 100.12 feet to an iron pipe found on the *westernmost boundary line of the Schultz Tract*; thence...

6.4.3.4 *Directions—Significant Figures* Not all descriptions with bearings employ the same level of precision in reporting the direction of a course. Some will contain a reference to a quadrant with only full degrees. Some will add minutes to those degrees. It is only relatively recent instrumentation and methodology (at least, in comparison to the multimillennial age of the surveying profession) that seconds of arc have been added to describe the direction of a line.

Early surveying practice on the North American continent was capable of establishing and reporting full degrees of arc, but the minutes of arc were interpolated between the compass lines marking full degrees. The development of dividing engines that could consistently and accurately mark finer divisions of a circle allowed surveyors to read minutes directly, requiring interpolation only of seconds. These indirect readings were of course subject to human error, but knowing the limitations of instruments contemporary to the original survey of a tract tells us more about the significant figures in the recorded angular measurements. This understanding of instrumentation and practice can also help us to ferret out possible scrivener's errors. We should investigate further if we find an angle recorded as "twenty six minutes" in a deed written when angles were reported only to the nearest quarter of a degree.

6.4.3.5 Distances   Just as with angular directions or bearings, the measurement of distances has evolved over the years, creating differences between distance reported in the written record for the same line, and between an early reported distance for a single line and its later measured distance on the ground. We must again recognize that accuracy is paramount, rather than reliance on the higher precision of modern equipment that allows us to measure much differently from the days of chains with links that only allowed direct measurement to the nearest full link (equivalent to approximately eight inches). We must always remember that accuracy is related to the intent of the real property transaction, and that pure mathematics cannot establish that intent.

The history of a tract and its creation is just as important in deciphering distances as it is for determination of directions. Was the parcel paced off by the original grantor; was it measured by counting the revolutions of a wagon wheel; was it created more recently and measured with an electronic distance measuring device?

6.4.3.5.1 Line Segment   We use the term *line segment* to mean the line or curve described by a single course of a description. Each line segment is described completely and separated from all other line segments by specific punctuation and phraseology that is not used in any other part of the narrative. This process divides the long and complex narrative into digestible bites, allowing the reader to concentrate on each line segment along with the termination point of that segment. Each boundary line segment in the body of narrative begins with the key word *thence* and ends with a semicolon (";"), even though these may appear in different lines of the description, depending on format that may be established by particular clients or agencies for whom the description is being written. In the absence of a specific format dictated by a regulatory agency or other unusual conditions, the clearest descriptions of line segments follow a programmed form. In the following example, the requisite key words are italicized.

thence, (general direction) *along* (boundary identifier), (specific direction), *a distance of* (specific distance) *to* (identify corner monumentation);

or

(general direction) *along* (boundary identifier), (specific direction), *a distance of* (specific distance) *to* (identify corner monumentation)*; thence*

Complex line segments, multiple line identifiers, corner information, and other expansions of this standard format will be discussed later.

Examples of line segments:

... ; thence, in a southerly direction along the westernmost right-of-way line of Kings Highway, South 15 degrees 32 minutes 15 seconds West, a distance of 325.62 feet to a $^1/_2$-inch-diameter iron rod set on the northernmost boundary line of Lot 23, Kings Estates; thence ...

... ; thence, northeastwardly along the southernmost boundary line of the Smith Tract, North 58 degrees 15 minutes 23 seconds East, a distance of 100.12 feet to an iron pipe found on the westernmost boundary line of the Schultz Tract; thence ...

These examples contain several key phrases that will be discussed later. Notice that even in the absence of a sketch, the reader can easily picture the orientation of the line segment discussed. All data presented between the key word *thence* and the terminating punctuation (";") pertains to one and only one line segment. Nowhere else in a land description should the key word *thence* be used except to initiate a line segment. The punctuation ";" is likewise reserved to indicate the termination of a line segment and is not used for any other purpose in the description.

6.4.3.5.2 Segment Length The line segment phrase, even when iso-lated by the key word *thence* and the punctuation ";", may have several distances associated with it. Curved lines will have radii, and corner monumentation may have several dimensions associated with it. The key phrase "a distance of" is used to identify that dimension that is the total length of the boundary line segment. The length should immedi-ately follow the key phrase. The example line segments are repeated here with the key phrase in italics and the segment length underlined.

... ; thence, in a southerly direction along the westernmost right-of-way line of Kings Highway, South 15 degrees 32 minutes 15 seconds West, *a distance of* 325.62 feet to a $^1/_2$-inch-diameter iron rod set on the northernmost boundary line of Lot 23, Kings Estates; thence ...

...; thence, in an easterly direction along the southernmost boundary line of the Smith Tract, North 88 degrees 15 minutes 23 seconds East, *a distance of* <u>100.12 feet</u> to an iron pipe found on the westernmost boundary line of the Schultz Tract; thence ...

6.4.3.5.3 Describing Curves

Straight boundary lines are defined by the location of their end points. But many boundaries are not straight lines. Natural boundaries, such as rivers, form one category of non-straight lines. Natural lines cannot be expressed in geometric terms and will be discussed in section 6.6.1.6.5 under "Wandering Topographical Calls." Curved lines, such as circles and spirals, can be expressed in geometric terms, but a good description provides information completely defining the line rather than merely providing a means of recovering the end points of the curved line segment.

Circular curves (curved lines with a constant radius) are the most common form of curved lines used to define real property boundaries. It is not sufficient to simply report the radius of a circular curve. The relative position of the radius to the line of travel (in the description) must be expressed as "a curve to the right" or "curving to the left" before the balance of the curve data is presented to bring the reader to the end point of the line segment.

The difficulty in expressing the additional data need has been the subject of considerable debate. Presentation of such data on the survey plat and, therefore, in the narrative of a description has been inconsistent. A resolution through design of rights-of-way is to restrict, as much as possible, the mixing of curved and straight lines to a tangent relationship. But we must still address existing conditions such as non-tangent lines and "broken back" curves that do not share a common radial line. The long chord bearing and distance between the end points of a curved segment is the preferred method of fixing the end points and is not reliant on the relationship of the curve radii with the connecting line segments. Some state and local regulations have begun the practice of requiring chord bearings and distances as well as radii, arc lengths, and direction of curve for all circular curves.

6.4.3.5.4 Tangent Curves

Some older descriptions were written from the point of beginning for a number of courses, then stopped and returned to the point of beginning to run in the opposite direction, all in an effort to avoid describing difficult courses, generally along curves or certain topographic features. Such a procedure creates an open description that cannot be mathematically closed. But, currently, in the vast majority of cases, and under ideal circumstances, the description of the end point of one line segment is the beginning point

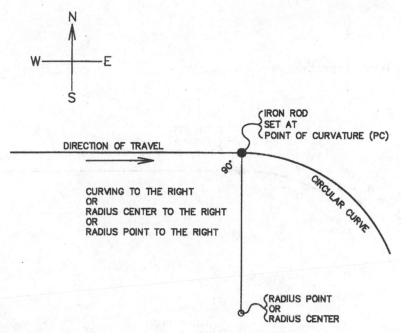

FIGURE 6.4 No matter which wording option is chosen, language should be consistent and clear throughout the description.

of the next line segment. In the case of circular curves, the transition from a straight line to the curved line can have a special relationship described by the key phrase "point of curvature" (abbreviated PC). This phrase is only used when the radius point or radius center of the curved line segment is positioned *at right angles* to the straight line segment leading into that curve. In the drawn example in Figure 6.4, the "line of travel" is from left to right.

The narrative for such a condition takes the form of:

> . . . to an iron rod and a *point of curvature*; thence, continuing in an easterly direction and following a line curving to the south and a radius of (radius) feet, a distance of (length of arc) feet to . . .

Many description readers are directionally challenged, and for that reason it is often easier for them to understand a narrative for this same curve with the following alternative language:

> . . . to an iron rod and a *point of curvature*; thence, continuing in an easterly direction and following a line curving to the right with a radius of (radius) feet, an arc distance of (length of arc) feet to . . .

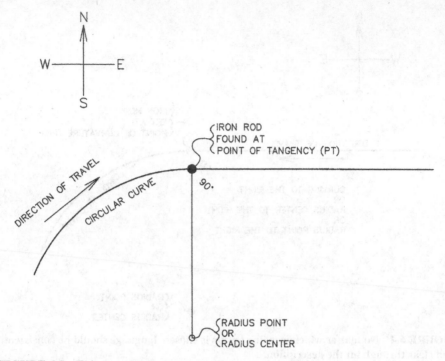

FIGURE 6.5   **This described course runs from a curved line segment to join a straight line segment.**

This form guides a reader along the described line in the most direct terms, so that looking at a survey or map is not immediately necessary to understand it.

The key phrase "point of tangency" (abbreviated PT) is used only when a circular curved line terminates at the beginning of a straight line segment and a line drawn from the radius center to that termination point forms a right angle with the straight line. In Figure 6.5, the "line of travel" is from left to right.

The narrative for such a condition takes the form of:

... feet to an iron rod and a *point of tangency*; thence, in an easterly direction, (bearing), a distance of (length) feet to ...

Compound curves and reverse curves are transitions from a curve of one radius center to another a curve having a different radius center. The radius centers in both cases must fall on a single line drawn from the radius centers and through the tangent point where the curves meet.

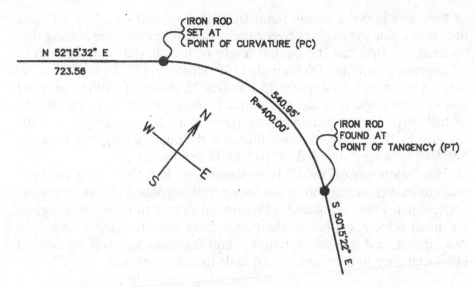

FIGURE 6.6 Example of a curved line between two tangential lines.

The use of tangent curves in the narrative portion of a description has two major disadvantages.

1. This method cannot be applied to configurations that are *not* tangential.
2. It can introduce computational anomalies because of significant figures.

Computations using the tangent definition of the transition from straight line to curved and back again will develop inconsistencies concerning the bearings of the straight lines.

A narrative for the sketch in Figure 6.6 could be developed using the tangential phrasing as follows:

...North 52 degrees 15 minutes 32 seconds East a distance of 723.56 feet to *a point of curvature;* thence, in an easterly direction following a curved line having a radius of 400.00 feet and the radius center to the south [or "a curved line to the right having a radius of 400.00 feet"], a distance of 540.95 feet to *a point of tangency;* thence, in a southerly direction, ...

Closure calculations of the example would disclose a discrepancy between the central angle of the curved section and the reported arc length. From the bearings presented, we can determine the direction of the lines into the radius point from the PC and from the PT, and then determine the angle between these two computed lines. Using this method, we find that the central angle of the curved line segment is 77 degrees 29 minutes 06 seconds (180 degrees – [52 degrees 15 minutes 32 seconds + 50 degrees 15 minutes 22 seconds]). When we plug this central angle into the geometric formula to compute the length of the curved line segment (using [central angle] = [Length × 180] divided by [Radius × Pi]), we find that the curve's length is actually 540.9473484 feet, rounded off to 540.95 feet on the plat.

This "rounding-off error" is accumulative. Parcels having multiple tangent curves or very long distances will introduce computation inconsistencies that can cause a theoretical failure to close. But, again, we must remember that mathematics does not drive the intent of a description, and that measurements and computations fall far behind other evidence in the hierarchy of calls in a description.

6.4.3.5.5 *Chord Bearing and Distance* A more exact method of describing circular curves has been gaining popularity. The chord bearing and distance method of describing circular curves can be used for any curved line segment. The radius center and straight line need not have a special orientation to each other. The computations used to "close the traverse" based on the chord bearing and chord distance are no more vulnerable to a rounding error than are any other straight lines in the description, and such closures are less problematic than with descriptions lacking these two elements.

We now add chord distances (CDs) and chord bearings (CBs) to the previous sketch, resulting in Figure 6.7.

The narrative for this line segment will now read:

> . . . North 52 degrees 15 minutes 32 seconds East a distance of 723.56 feet to a (describe point by its record and/or its physical monument) *and point of curvature;* thence, in an easterly direction following a curved line having a radius of 400.00 feet and the radius center to the south [or "a line curving to the right having a radius of 400.00 feet"], a distance of 540.95 feet, that same line having the chord bearing of South 88 degrees 59 minutes 55 seconds East and a chord distance of 500.66 feet, to a (describe end point); thence, in a southerly direction, . . .

Using the chord bearing and chord distance method of describing circular curves eliminates the need to specify tangent relationships. Circular curve line segments that do not have tangent relationships can be described as easily as those that do. Of course, in order to

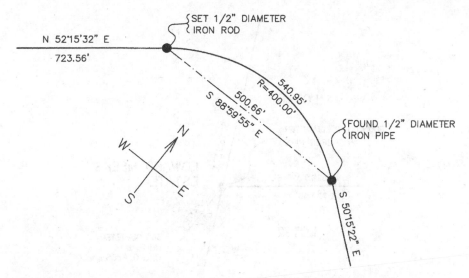

FIGURE 6.7 Chord data adds a means to check the orientation of the curved line.

describe curved lines using this method, the survey plat must present the necessary data.[7]

6.4.3.5.6 Accessories The term *corner accessory* is a specific PLSS label, but its meaning is universal throughout surveying practice in the United States, whether PLSS or colonial.

Accessories are physical evidence in the vicinity of a corner or a corner monument. The measured, recorded relationship between the accessory and the corner serve to recover the position of the corner monument if it is destroyed or removed, or to witness a corner that is inaccessible for purposes of setting a stable permanent corner monument. Suitable objects for use as accessories in the PLSS are bearing trees and other natural objects, mounds of stones, pits, and memorials. In each of these instances, the measured or calculated distance and direction between the true corner and the referenced accessory is recorded in the field notes and memorialized on the survey plan and in the description.

The term *memorial* has a slightly different meaning in the PLSS and colonial systems of surveying. In the PLSS, this is a surface marker set where no trees or other bearing objects are available to tie to the tract corner. In colonial surveying, however, this term more frequently means a subsurface marker set below the corner monument, to preserve

---
[7]See Chapter 5.

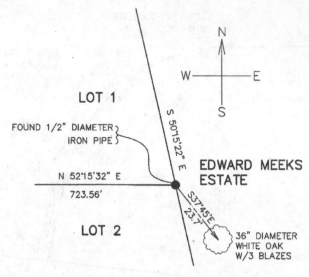

FIGURE 6.8   Survey showing both a corner marker and an accessory to the corner.

the corner if the surface marker is lost. In early days, the object serving as the memorial was generally whatever happened to be on hand, sometimes a unique stone and sometimes an empty whiskey bottle. But it would always be carefully included in the field notes. In instances where a survey was closer to home, the subsurface memorials could be prepared ahead of time and might be ceramic jars or rings, as were set for the first triangulation survey of the East Coast under Ferdinand Rudolph Hassler in the early 1800s.

No matter the name given to the accessory and no matter the type of accessory, any such physical evidence helps to perpetuate the position of the property corner. The recorded description of the tract, therefore, serves to give broader public notice of that evidence. Any accessory, witness, or reference monument, like those shown in Figure 6.8, should be fully described as to material, size, and orientation (offsets or a bearing and distance tie) to the true corner.

> ... to a point on the southwesterly line of lands now or late of Edward Meeks, evidenced by a 36-inch diameter white oak with three blazes found South 37 degrees 45 minutes East, 23.7 feet from said point; thence ...

> ... to a point in the northerly right-of-way line of the aforementioned Quail Way, evidenced by a $^3/_4$-inch capped rebar marked "Feldman" found 0.45′ south and 0.87′ east of said point; thence ...

> ... to a point in the centerline of Stormy Run, passing over a concrete monument set on line 250.00 feet from the origin of this course; thence ...

6.5 ELEMENTS OF THE DESCRIPTION

Beyond the caption and body of the description, other elements assist in capturing the intent of the parties. It is the duty of the description writer to use words to fully explain that intent rather than create a situation in which later readers will be tempted to force the intent to follow the written words.

6.5.1 Qualifications (Additions, Subtractions, Reservations)

The terms of the contract expressed in the deed may include qualifications as to which party will benefit from certain conditions or which will give something up. These qualifications can be additions to the rights already described in the deed, or they may be subtractions from them.

6.5.1.1 Augmenting Clauses *Augmenting clauses* add benefits to a property, such as when a tract is conveyed *together with* certain other rights. This can be anything from an access easement over an adjoining tract to the right to share the dock along the common line with another lakefront landowner. Besides easement rights, augmenting clauses may convey other privileges, such as the right to collect royalties on a timbering lease or the right to a certain amount of irrigation water on a regular basis. Sometimes the language of augmenting clauses paints a clear picture of the kind of arguments leading up to the formality of setting down the rights in words.

> Together with the free and common use, right, liberty and privilege of the aforesaid driveway apron extending the full width of the garage in the rear of the said building, between the vehicular entrances to the said garage facing it at the rear of the building on the lot adjoining to the Northwest, as and for an automobile driveway and passageway for driving from and out of said garage at all times hereafter, forever.

6.5.1.2 Encumbrances *Encumbrances* are conditions to which a property is subject, such as easements or restrictions on use. Easements in favor of another tract are most often introduced with the words "Subject to . . .".

> Subject to a ten-foot wide easement to Public Service Electric Company for the purpose of maintaining its power poles, being more particularly described as follows:

Subject to the rights of the public in and to so much of the described premises described herein as lie within the bed of Walnford Road.

Subject to a 25' wide "perimeter patrol road" easement for the benefit of the County of Monmouth for the maintenance and patrolling of the stream valley. Said easement runs along the property lines and/or wood lines between the subject property and the lands of the County of Monmouth, Lot 10 Block 49 as set forth in Deed Book 5401, Page 702.

Subject to a private right of access, ingress, and egress to benefit adjoining Lot 12, Block 32 as described in Deed Book 4422, Page 95.

Encumbrances may also come in the form of covenants or conditions. These can trigger future rights of the grantor, such as when a covenanted prohibition against using the existing building as a tavern is broken, returning the land to the grantor. The second example could be restricted further by adding specific language.

Subject to a private right of access, ingress, and egress to benefit adjoining Lot 12, Block 32 as described in Deed Book 4422, Page 95. Any change in use of Lot 12 from residential is subject to review and approval of the Grantee of the present conveyance and transaction. Use of said private way is meant only for the current dwelling located on Lot 12 and shall not be extended to service additional lots developed from any division or reconfiguration of Lot 12.

Of course, an existing encumbrance cannot be modified by anyone other than the two parties benefitting and encumbered by it. If the first version of this easement was the condition under which the grantor of the servient estate took title, the new grantee cannot make any changes to the contractual agreement that clause represents without negotiating and formalizing a written agreement in the form of another deed with the dominant estate owning the easement rights.

At times, grantors wish to keep rights to part of the conveyed land for their own future use, and the term used to describe this situation is a *reservation* of those rights, resulting in another means of encumbering a property. For instance, a grantor selling the part of his land that fronts on a county road and effectively leaving himself landlocked may reserve a right to cross the land he has just deeded away in order to access the county road. Terms of a reservation may include how many people may exercise that right, at what times of day, using what mode of transportation, or any other conditions agreed to by the original parties to the transaction. These reservations are newly created rights, and may be for any variety of purposes, from the right to farm part of the conveyed premises to the right to drill oil.

In the following example, the Grantors conveyed various parts of their tree farms to become part of the County Park System, with certain conditions attached to that conveyance:

The Grantors herein expressly reserve unto themselves and their successors and assigns the rights and easements set forth below on, over, under, and across the Parklands for the benefit of the lands and premises remaining to them after the within conveyance and any other lands and premises, now owned or hereafter acquired adjoining the Parklands or adjoining lands within the County Park System which adjoin the Parklands (collectively, the "Nurserylands"):

The deed next contained various paragraphs outlining very specific rights that the Grantors had made sure they would still be able to exercise after the conveyance of what was to become Parklands. Among others, these reserved rights included entering the Parklands to spray the stream area for gypsy moths, (with certain care and notification requirements specified), entering the parklands to maintain existing irrigation pipes for their tree nursery (in newly created irrigation pipeline easements), and the right to hunt deer on the Parklands for the specific purpose of herd management.

6.5.1.3 Exceptions *Exceptions,* sometimes also termed *deletions* or *subtractions,* prevent transfer of interests from conveyance of the entire tract that has just been described. It may be that a conveyance out of the larger tract had previously occurred but no new survey provides a description of the lands remaining and available for conveyance in the present deed. In such instances, the full undivided tract may be described, followed by a phrase something to the effect of:

Excepting from the above described premises a tract previously conveyed to John Smith on June 15, 1990 and recorded in Deed Book 1234, Page 567.

There may also be a subdivision of a tract and simultaneous conveyance of its parts that must be addressed in separate descriptions, such as in the following example:

Excepting from the above described tract a strip of land along the north side of First Avenue to be dedicated this day to Morris County for road widening purposes, described as follows:

Reservations and exceptions are two completely different legal animals but unfortunately there are numerous examples of descriptions employing both terms in the phrase "excepting and reserving. . . ." An exception refers to a right that already exists and is being kept by the grantor, while a reservation refers to one that is being created for the first time with the current documents. To illustrate the distinction, think of the common statement, "I reserve the right to change my opinion when additional information is made available to me." While there is no current change in the speaker's opinion, a future right has been created for the first time with this statement.

The same concept applies to written land descriptions. While it is often difficult to distinguish between an exception and a reservation in a description, and the words *excepting* and *reserving* are not conclusive in determining which is meant, it is the character and the effect of the provision in which these words occur that will shed the most light on the intent. If the intent is to keep some part of the grantor's former estate, excluding it from the grant, then this is an exception, being a part of the estate not granted away. A reservation, however, is something taken back from whatever is clearly granted, such as a reservation for right-of-way, water, or light.

Take, for example, the following description, which became the subject of litigation more than 80 years after it was signed, sealed, and memorialized in the County's Hall of Records:

> Excepting and reserving a right-of-way sixteen (16) feet in width along the southerly line of the above described premises running easterly and connecting with certain other rights-of-way described in deeds recorded in the Bergen County Clerk's office in book 776 of deeds page 258 book 896 of deeds page 148 and the deed to Rathburn and wife above referred to said right-of-way to be for the lawful use and benefit of all owners of the premises herein described as well as owners of other lands of the grantor situate west of said premises.

The plaintiffs' initial argument was that this statement held back the fee title to a 16'-wide strip of land from the original conveyance, based on the opening term, *excepting*. When we look at the rest of the paragraph, however, we see that there is a specific purpose established for this strip of land, a right-of-way that will continue to serve the remaining (and landlocked) lands of the grantor and whomever his later grantees of those remaining lands will be. Title is comprised of the entire bundle of interests and not restricted as to use; therefore, this is a reserved easement right for a particular use rather than an exception from the transfer of full title and ownership. The only "exception" is from unencumbered status. The phrase "excepting a right-of-way" is generally construed to mean an easement rather than fee title in that land. Combining both "excepting" and "reserving" in the same sentence serves only to confuse readers unfamiliar with the terminology and precepts of real property law.

6.5.2 Closing and References

The final paragraphs of a description provide the source of the information that was presented earlier within the caption and body. For a description based on a survey, provide adequate information for a future reader to locate the document.

> The above is intended to describe all that parcel of land shown on a survey by John Doe, PLS, Surveying Consultants, Inc., Wilmington, Delaware, dated May 15, 2008 and revised to October 12, 2009, marked as Project #45-HTS.

> The above description was written pursuant to a plan entitled "Boundary Survey and Minor Subdivision of Lands of Winston Selarek, Zion Road, Tranquility, New Jersey," filed in the Warren County Clerk's Office on May 20, 1902 as Map No. 254.

A reduced copy of the survey can also be attached to and made part of the description, and in such instances that fact should be included in the closing language.

Descriptions prepared by attorneys or other nonsurveyors more frequently reference the prior owner of the described parcel than a survey, past or present. It is common to find closing references along the line of:

> Being the same tract conveyed to Mary Smith by Thomas Smith in Deed Book 123, Page 45 on January 10, 1997.

Such references do provide valuable information to surveyors and others who need to trace the chain of title backwards for various reasons, whether to settle ownership questions or to find the source of a discrepancy between adjoining deed descriptions.

6.6 PUNCTUATION AND LANGUAGE

The proper use of punctuation is important in any written document, and within a description it is no less so. A misplaced or missing comma or semicolon can alter the meaning of a phrase and/or introduce ambiguity.

The distinction between commas and semicolons seems to be a fuzzy area for many people, yet inclusion or omission has in fact been the pivotal point of at least one lawsuit. When determining if a warranty in a contract for sale of land and its appurtenances covered the orchards and related equipment *or* the real property *or* both, the court in *Sackman Orchards v. Mountain View Orchards* (784 P.2d 1308, Court of Appeals of Washington, Division Three, 1990), had this to say:

> The omission of a semicolon has fomented confusion from clarity. The focal issue is whether it was intentionally omitted or is an isolated grammatical/typographical faux pas. The omission has allowed creative lawyers to obscure the clear intentions of the parties.[8]

The disputed contract for sale read as follows:

> PROPERTY SOLD: On the terms and subject to the conditions set forth in this Real Estate Contract, Seller agrees to sell to Buyer and Buyer agrees to purchase from Seller the equipment stated in Schedule 1 attached hereto and incorporated herein; the 1982 fruit crop[;] and the following legally described real property in Douglas County, Washington, with title free of encumbrances and defects, except as set forth on and subject to Schedule 2 attached:

In addressing the question of what was intended to be covered by this contract, the court noted:

> Had the series, *i.e.*, equipment; crop and real property, been separated by a semicolon after crop – or a comma used after equipment and again after crop – there would be no ambiguity. The omission of either resulted in filing this action, this appeal, and the differences in interpretation.[9]

In a footnote to its opinion, the court also provided a lesson in punctuation:

> A semicolon is used to join two or more clauses, which are grammatically complete, to form a single compound sentence. W. Strunk & E. B. White, *The Elements of Style* 5, 6 (3d rev. ed. 1979). A "serial" comma is used after each term, except the last, in a series of three or more terms with a single conjunction, *e.g.*, French, German, Italian and Spanish. W. Strunk & E. B. White, at 2. Some grammarians instruct that a comma is used in a series immediately before the single conjunction to avoid ambiguity or for stylistic uniformity, *e.g.*, French, German, Italian, and Spanish. H. W. Fowler, *A Dictionary of Modern English Usage* 588 (2d ed. 1965).[10]

---

[8] 784 P.2d 1308 at 1309 [footnote omitted].
[9] Ibid.
[10] Ibid.

6.6.1   Key Words or Phrases

When we speak of "key words" or "key phrases," these represent language that provides insight to the intent of the description. All who read a document must understand it consistently; proper use of key language helps to build a clear, concise, coherent description that will not lead to future arguments over its meaning.

In the pursuit of the perfect description, there are a few terms of which we should wary. Some of these represent lazy document preparation, some exemplify misuse of the English language, and in each case alternative phrasing better serves the immediate users of the description as well as all later readers of the written record.

6.6.1.1   "Point"   We often find descriptions in which the boundary lines run "to a point" with no further designations. While the definition of a "point" in geometry may be an exact location, it is neither precise nor accurate in establishing the end of a boundary line; the ending point is merely one of the series of points creating that line. A "point" lacks dimensions, has no color or shape, and provides no evidence of the intended termination of a line. Early descriptions in our country included vast amounts of information regarding the topography, physical features, and adjoining landowners, but over time much of this information was eradicated from the public record, whether for the sake of brevity or due to the ignorance of the rewriter of the description. Good descriptions do not terminate courses with the phrase "to a point", and instead provide a more complete and through description of what physical or record evidence exists in these locations.

6.6.1.2   "More or Less"   The phrase "more or less" refers to an approximation in which slight or unimportant inaccuracies in quantity are considered inconsequential to the overall quantity. It is often used in the recitation of area at the end of a description, which is not generally problematic except when this runs contrary to the intended conveyance when "exactly" a certain number of acres were to be conveyed, or when significant figures come into play. Areas in acres reported to four or five decimal places should not be recited as being "more or less," particularly if the number of square feet in the tract is reported at the same time. Area of tracts that are bounded by natural features such as bodies of water certainly can be cited as "more or less." Those natural boundaries often cannot be measured as precisely as monumented and established lines and will be reported as "more or less" lengths.

Another use for "more or less" is to follow the general course of a feature or proceed in a general direction. For example, a stone fence row may be in the general area of the boundary. If the stone row is meant to be the line of demarcation between two properties, then the line of the deed description should run along that feature, and a bearing supplied based on the beginning and ending points of that stone row will only be "more or less" along the center or face of that stone fence row. Here is where the language of the description must be explicit. If it is the stone row that is intended to control the property line, then the line runs:

> . . . along the stone row, *more or less* along a bearing of *(fill in the blank)*.

However, if the stones were merely tossed to the side of the field by farmers after the tracts were subdivided, then the stone fence row they create does not control the boundary. In such a case, the description should read:

> . . . along a bearing of *(state the bearing)*, also running *more or less* along the center of a stone fence row.

These two statements are quite different in meaning, even though they share terms and references in common. If there is any ambiguity in the final written document, a court could decide its meaning according to the hierarchy of calls as the best evidence of the original intent, which may or may not coincide with the true but unexpressed intent.

Under no circumstances should the phrase "more or less" be used to cover up or gloss over an unresolved problem or ambiguity. The description writer should be working to preserve and clarify the terms of a conveyance, not to protect sloppy or incomplete work. Ask questions, research more written records, perform more fieldwork—but do not pass difficult situations on to the next reader or user of the document. The use of "more or less" should be used sparingly and only when specific conditions such as those discussed in this section warrant such lack of certainty in the description.

6.6.1.3 *Monumentation and Boundary Designations: "To"* The size, shape, material, and other identifying attributes of the monumentation set or found at boundary corners is vital in the recovery of real property boundaries. Often, key words and phrases express the intention of the parties involved in a real property parcel creation or transfer.

The key word "to" defines the termination of a course, identifying a record or physical monument that establishes where that line ends. When used following the line segment length, "to" affects application of the hierarchy of calls, as we will show in the following examples. It is the identifying feature following the key word "to" that will take precedence over the reported length of the segment. We should remember that "to" in the context of a property description is generally an exclusionary term, meaning that the line ends at the monument called to at the end of that line. However, in terms of a three-dimensional monument rather than a record monument, "to" runs to the center of that monument unless specifically stated otherwise. A call "to" a 32″ diameter copper beech tree runs to the center of that tree, and similarly a call "to" a stream or a road will be interpreted as running to the centerline unless additional language clearly excludes anything beyond the bank of the stream or the sideline of the road.

The italics in the following example line segments represent application of the key word "to" and its role in preserving evidence of a boundary.

> . . . ; thence, in a southerly direction along the westernmost right-of-way line of Kings Highway, South 15 degrees 32 minutes 15 seconds West, a distance of 325.62 feet *to a $^1/_2$-inch-diameter iron rod set on the northernmost boundary line of Lot 23, Kings Estates*; thence, . . .

> . . . ; thence, in an easterly direction along the southernmost boundary line of the Smith Tract, North 88 degrees 15 minutes 23 seconds East, a distance of 100.12 feet *to an iron pipe found on the westernmost boundary line of the Schultz Tract;* thence . . .

> . . . ; thence, in an easterly direction along the southernmost boundary line of the Smith Tract, North 88 degrees 15 minutes 23 seconds East, a distance of 100.12 feet *to a point on the westernmost boundary line of the Schultz Tract evidenced by a found iron pipe;* thence . . .

In the first example, the hierarchy of calls sets the boundary line of Lot 23 as the terminus of the line segment, even if this requires the line segment to be longer or shorter than the reported 325.62 feet; the intent of running to a known boundary line is more important than the measured distance to that line and marker. At the same time, if an iron rod is found in the vicinity but it does not happen to be on the boundary of Lot 23, it will be given less weight in the search for evidence of the intended course.

In the second and third examples, the boundary line of the Schultz Tract is the termination of that line segment, even if this requires the line segment to be longer or shorter than the reported 100.12 feet, or an iron pipe is found at a location other than on the Schultz Tract boundary.

The meanings of these descriptions are completely changed if the examples are amended to read as follows:

... ; thence, in a southerly direction along the westernmost right-of-way line of Kings Highway, South 15 degrees 32 minutes 15 seconds West, a distance of 325.62 feet *to a ¹/₂–inch diameter iron rod;* thence ...

... ; thence, in an easterly direction along the southernmost boundary line of the Smith Tract, North 88 degrees 15 minutes 23 seconds East, a distance of 100.12 feet *to an iron pipe;* thence ...

Now the hierarchy of calls leads us to interpret the iron rod and the iron pipe as the respective terminal points of the courses, rather than the line of another boundary. The reported lengths of the lines still are of lesser weight than the markers called for in the descriptions.

Too often, we find that these examples would be amended to read:

... ; thence, in a southerly direction along the westernmost right-of-way line of Kings Highway, South 15 degrees 32 minutes 15 seconds West, a distance of 325.62 feet; thence ...

... ; thence, in an easterly direction along the southernmost boundary line of the Smith Tract, North 88 degrees 15 minutes 23 seconds East, a distance of 100.12 feet; thence ...

or:

... ; thence, in a southerly direction along the westernmost right-of-way line of Kings Highway, South 15 degrees 32 minutes 15 seconds West, a distance of 325.62 feet *to a point;* thence ...

... ; thence, in an easterly direction along the southernmost boundary line of the Smith Tract, North 88 degrees 15 minutes 23 seconds East, a distance of 100.12 feet *to a point*; thence ...

A great deal of boundary evidence has been purged from the descriptions in each of these versions. The reader is left with only the reported length and direction of the line segment to define the termination of the boundary line segment. All of the higher-ranking corner attributes vital to the rules of interpretation under the hierarchy of calls are absent. Future readers of the document have been deprived of the primary means of identifying the real property corners.

6.6.1.4 Boundary Designations: "Along," "Generally Along", "On and Along"

In some contexts, the term "along" merely represents a relationship lengthwise of, at or near, but not necessarily touching at all points or even having any contact at all. Therefore the intent must be clearly expressed. In writing a description, ask yourself if the course is running along a line or along a strip, such as one defined by a line of subdivision, road right-of-way line, or mean high water line of a lake. "Along" is suitable for a line. But sometimes the "line" has width because it is a physical object—maybe it is a fence or a stream. If the physical object is intended to control the location of the boundary, "on and along" may better accommodate the twistings and turnings of that object as it travels between its beginning and ending points. If the physical object is in close proximity to but not controlling the boundary, the bearing of the line may be referenced as "generally along" the physical object.

We offer one final caution in using the term "along" in relation to a strip such as a road or stream. If unqualified, "along" in this context will be taken to mean along the center of the road or stream.

The key word "along" can also tell us that a described boundary line is coincident with another line. A line segment described as being "along" an identified boundary eliminates any possibility that the description could be construed as having a gap or overlap.

Consider the sketch in Figure 6.9. If the land description contained the segment "... thence, in an easterly direction *along the northernmost right-of-way line of 33rd Street*, North 89 degrees 36 minutes 23 seconds East, a distance of 200.00 feet to the westernmost boundary line of Lot C ...", there would be no doubt that Lot B fronted 33rd Street. Even if a subsequent survey determined the bearing of 33rd street to be North 88 degrees East, it could not be argued that there is a sliver of land between Lot B and the street right-of-way. The land description clearly stated that the southernmost boundary of Lot B is the same line as the northernmost right-of-way line of 33rd Street.

The example line segments are repeated here with the key phrase in italics.

> ...; thence, in a southerly direction *along the easternmost right-of-way line of Kings Highway,* South 15 degrees 32 minutes 15 seconds West, a distance of 325.62 feet to a $^1/_2$-inch-diameter iron rod set on the northernmost boundary line of Lot 23, Kings Estates; thence ...

> ...; thence, in an easterly direction *along the southernmost boundary line of the Smith Tract,* North 88 degrees 15 minutes 23 seconds East, a distance of 600.12 feet to an iron pipe found on the westernmost boundary line of the Schultz Tract; thence ...

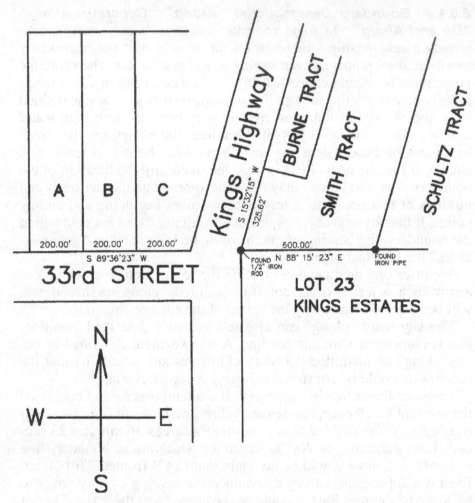

FIGURE 6.9   Boundaries along road rights-of-way.

Because of the legal presumption that properties situated along roads and highways include the land to the centerlines of these features, a description "along 33rd Street" or even "along the northern right-of-way line of 33rd Street" does not exclude the land between the right-of-way and the centerline of 33rd Street. If such exclusion is intended (in the absence of a clear fee conveyance of the roadbed to another entity), an augmenting clause at the end of the recited description courses must clearly state: "Excepting from this conveyance the title to the land between the centerline and northerly right-of-way line of 33rd Street," or words to this effect. Rather than transferring title to the grantee, this either acknowledges prior conveyance of the roadbed (probably but not always to a governmental entity) or it reserves the roadbed to the grantor (or the grantor's heirs, successors, or assigns). When tracing the chain of title, these distinctions and clarifications can later make a significant difference in determining the rights of owners abutting a publicly traveled way, particularly when it comes time to widen the road right-of-way and appropriate compensation must be established based on the amount and kind of interest held by private landowners.

6.6.1.5 *Providing a Complete Record*   Often, a boundary or a corner will have multiple aspects that must be documented to provide a complete record. A corner may be monumented by an iron pipe, associated with a witness tree, and reported to be at a certain geographical location. A boundary line may begin as the boundary for one adjoiner and then continue as the boundary for a second. Each of these facts provides additional evidence of the corner, building a strong means of preserving its location.

6.6.1.6 *Careful Use of Language*   The descriptions we write will become part of the chain of title, a means of preserving intent and evidence of boundaries. The choice we make between one word and another may affect future understanding of where a boundary lies. Understand the full meaning of words before employing them, and use them properly.

6.6.1.6.1 *Relationship to a Reference Line*   When describing a baseline or reference line for a strip of land that is not interior to that strip, the width of that strip is on *each* side of the described line. The width may be equal on both sides of the described line or it may be different, but there is a width on each side rather than on just one side of the described line. If only on one side, then the introduction to that description should state, for example, that the line about to be described is the northerly line of a 25' wide strip. Too often, we see the word *either* when the baseline, centerline, or reference line is interior to the strip of land. However, *either* means that there is a choice: the stated width can either be on this side or on that side of the baseline, but it can't be on both sides.

If our reference line is a centerline, then it has an equal width of land *on **each** side* of it or *on **both** sides* of it. It doesn't matter if the centerline is straight or curved: . . . *being an access right-of-way 15 feet wide on **each side*** [or ***on both sides***] *of the herein described centerline.* . . . Don't give the reader a choice of applying the width on one side or the other by providing a width "on either side" of the reference line.

A strip of land having a different width on each side of the described reference line requires us to state a geographic relationship as well as the width: *. . . being a right-of-way for utility purposes 15 feet wide on the southeast side and 25 feet wide on the northeast side of the herein described reference line. . . .*

6.6.1.6.2 *Relationship between Boundaries* The term *coterminous* (also *conterminous*) describes the relationship between two tracts that share a common boundary. Similarly, a boundary line described as *coincident with* another parcel's boundary means that the boundaries of those two tracts occupy the same place or position, and exactly correspond to and coincide with each other. But lines that are *contiguous with* each other may or may not be in full contact at all points. While in close proximity to each other, contiguous lines may merely be neighboring or adjacent but not fully coincident with each other.

Similar distinctions exist between the words *adjacent, adjoining,* and *abutting. Adjacent* features lie near or close to each other, but they may not actually touch. *Adjoining* features are in contact with each other, as opposed to merely being adjacent. *Adjoining owners* share some amount of common boundary, even if their ownership lines then extend beyond or angle away from each other. *Abutting* features touch each other in the same sense as *contiguous,* and are therefore also adjoining. *Abutting properties* have no land intervening between them.

Consider carefully before choosing the term that best describes the relationship intended.

6.6.1.6.3 *Relationship between Lines* As a matter of geometric and grammatical propriety, straight lines are *parallel with* each other when they lie in the same plane, extend in the same direction and are equally distant from each other at all points. However, when the intersection of two lines forms a 90-degree angle, those lines are *perpendicular to* each other, or alternatively, they intersect *at right angles to* each other. Curved lines lying in the same plane, extending in the same direction, and equally distant from each other at all points are *concentric with* each other.

6.6.1.6.4 *Geographic Orientation* There are times when we want to describe a general direction, one not tied to degrees, minutes, and seconds. Perhaps we are entering a curve and want to tell the reader approximately where we are heading before we provide all the details. Maybe we are describing a stream that doesn't run in a straight line.

We can describe these general compass-oriented directions in several ways. If we are going to use the description as an adjective, then we use the direction itself with the suffix *-ly: Thence **in a northerly direction** along a curve to the right* . . . But if we are using the term as an adverb, then we use the direction itself with the suffix *-wardly: Thence* **running southwestwardly** *along the various courses of the creek . . .*

The relationship between physical objects or lines utilizes the term *of: Thence along a line 16.5 feet* **southwest of** *and parallel with the centerline of the road . . .*

6.6.1.6.5 *Wandering Topographical Calls*

Not every course in a description fits in a nice neat geometric package. Rather than a straight line that heads in a specific direction for a specific distance or curved line that has a particular radius length and central angle, certain topographic calls wander, sometimes mildly and sometimes with significant variation. For courses that meander only a bit and travel generally in one direction, we can add language that provides the reader a little more guidance: *. . . thence running along the various courses of the center of Beden's Brook in a general northwesterly direction . . .*

If we are addressing a widely variable course, perhaps along the winding shoreline of a bay, we can also add a computed tie between the origin and terminus of the course in the form of a bearing and distance, just as we add chord information when describing a curved line. This practice also simplifies plotting the description by forming a closed polygon rather than one with unplottable sides, although, of course, the area of the polygon will not be the same as the area of the actual land parcel.

6.6.1.6.6 *Previously Mentioned References*

When we have provided a thorough explanation of the characteristics of a monument or cited a full deed reference and lot identification of an adjoining tract, we need not repeat that language fully in every subsequent mention of those features. If the full identification has just been provided, we can insert the word *said* in the next line referring to that same item. For instance, we can state that a boundary course runs "to the centerline of Anderson Avenue, 66 feet wide per County Ordinance 92-110," and the next course runs "thence, along *said* centerline. . . ." Use of the term *aforesaid* or *aforementioned* allows us to refer back to a previously mentioned or described feature, line, or adjoining owner without reciting the full lengthy description already provided earlier in the document. *Aforesaid* or *aforementioned* need not appear immediately after the referenced text, and several courses may separate the first mention and the later reference to it as *aforesaid*.

The key phrase "that *same* (point, line, or object)" or "*said* (point, line, or object)" is used to include additional information before the description moves on to another point or line. When the reference to the same point, line, or object does not immediately follow the first mention of that item, the correct phraseology is "the *aforementioned* (point, line, or object)."

In the following example, both *said* and *aforesaid* are employed to indicate departure from and return to a previously mentioned reference.

> Beginning at a railroad spike found in the centerline of Old Cranbury Road (66′ wide per Road Return) at the intersection of the southeasterly line of Lot 42, Block 97, lands now or formerly of Elizabethtown Water Company, and running; thence

1. Along the *said* centerline of Old Cranbury Road, South 20 degrees 58 minutes 31 seconds East, 50.32 feet to a point, being the most northerly corner of Lot 47, Block 97, lands now or formerly of Arthur Coe; thence

2. Along the northwesterly line of *said* Lot 47, South 69 degrees 01 minute 29 seconds West, 150.00 feet to a $^3/_4$-inch rebar found for a corner; thence

3. Along the southwesterly line of *said* Lot 47, South 10 degrees 58 minutes 31 seconds East, 50.77 feet to an angle point; thence

4. Along *that same line,* South 30 degrees 58 minutes 31 seconds East, 50.77 feet to a $^3/_4$-inch rebar found for a corner; thence

5. Along the southeasterly line of *said* Lot 47, North 69 degrees 01 minute 29 seconds East, 150.00 feet to a railroad spike found in the centerline of the *aforementioned* Old Cranbury Road; thence

6. Along the *said* centerline of Old Cranbury Road . . .

6.6.2 Construing Ambiguous Deeds

The term *ambiguous* in relation to deeds refers to a question as to the intent of the parties, either because the terms of the deed are not consistent with each other or because the terms of the deed do not seem to match what is on the ground. In such instances, the courts allow us to construe deeds, with the term *construe* defined in *Black's Law Dictionary* as:

> . . . To ascertain the meaning of language by a process of arrangement and inference.

The process of construing a document is called *construction,* which is further explained in *Black's Law Dictionary* as:

> The process, or the art, of determining the sense, real meaning, or proper explanation of obscure or ambiguous terms or provisions in a statute, written instrument, or oral agreement, or the application of such subject to the case in question, by reasoning in the light derived from extraneous connected circumstances or laws or writings bearing upon the same or a connected matter, or

by seeking and applying the probable aim and purpose of the provision. Drawing conclusions respecting subjects that lie beyond the direct expression of the term. The process of bringing together and correlating a number of independent entities, so as to form a definite entity.

This long explanation states the essence of why we need to write clear, complete, and concise descriptions: if something is not explicitly expressed in the document, then the description may be subjected to a process of "construction" by someone other than the drafter of the description or the parties to the original transaction. This leads to interpretation by whatever other best information that someone else may gather. It is always possible that the next "someone" will not bother or will not be able to gather the same evidence as you had in preparing your description. Either carelessness by our successors or destruction of evidence by the passage of time may serve to foil what the drafter of the description had believed was a clear representation of conditions.

This is not an invitation to wordiness or repetitiveness, but an admonition to be complete and clear, with properly used terms and avoidance of language that can be interpreted differently by different readers, either due to different (or lacking) background or to highly localized terminology. One very strong incentive to writing a good description is that in the face of perceived ambiguity, in a fee title conveyance the description will be construed or interpreted most strongly in favor of the grantee rather than the grantor, since preparation of the document presumably was under control of the grantor who could have said specifically whatever he or she meant, while the grantee merely accepts whatever terms are signed off by the grantor.

As with many rules, there are exceptions to this one as well, and the case of *United States v. 5.324 Acres of Land* (79 F. Supp. 748, U.S. District Court for the Southern District of California, Central Division, 1948) provides an illustration.

The federal government filed a Declaration of Taking against land owned by Mrs. Grace B. Severy, who sued when she was not compensated for the portions of the neighboring streets adjoining her tract, on the basis of her ownership to the center of the streets. But the United States argued that the metes and bounds description in its deed, which did not include those streets, was adequate to overcome the presumption under California statutes that landowners abutting streets own title to the center of those streets.

The court soundly rejected the United States' arguments, pointing out that the Constitution demands that the government at least be specific in designating property it wants to condemn rather than "be permitted to indulge in statutory presumptions which would not be necessary in this case if the Government had been careful in the first place to describe the streets in its original Declaration of Taking."[11]

The United States alone was in control of preparing the instruments for its condemnation proceedings, including the description to be used. "Mrs. Severy had no intent to voluntarily grant *any* of her land to the government, and had nothing to do with the metes and bounds description prepared by counsel for the Government who specialize in this type of procedure."[12] Had she been a *willing grantor*, then the description in the deed to the United States would have been interpreted differently and title would have been conveyed to the center of the streets, even lacking specific mention of that area.

Related to the idea of willing grantors of interests, descriptions for easement rights are generally construed in favor of the grantors, who are giving up their rights, rather than in favor of the grantees, who generally dictate the terms and location of the encumbrances they wish to impose on the land.

6.6.2.1 *Legislation and Court Rules*   One means of construing an ambiguous deed is the application of certain uniform approaches to the process, known as *rules of construction,* to decipher the meaning of the document. These rules of construction are generally found within a state's statutes or as *rules of evidence* within its court rules issued by a specific state's judiciary branch detailing evidence that may or may not be admitted to court, and for what purposes. While the passage of time has changed some of the language of these rules, their general approach to ferreting out the meaning of an ambiguous document and protecting the rights of grantees of fee interests and grantors of easements have remained unaltered. Note also that there are rules of construction established for many writings beyond deeds, but all such rules aim to provide a just and consistent determination of meaning.

One universal rule across the country is that that only evidence presented in a written instrument (meaning *intrinsic evidence,* rather than *extrinsic evidence*) shall be used to interpret an unambiguous deed. Like many rules, this one has exceptions, and these fall under the rules of construction. The first such exception to relying only on the contents of the written instrument to interpret that document arises when the validity of the instrument itself is in dispute. Closely related to this

---

[11]79 F. Supp. 748 at 761.
[12]79 F. Supp. 748 at 762.

exception is a second one that permits extrinsic evidence to establish illegality or fraud related to the creation or execution of the instrument of conveyance. Other allowances for extrinsic evidence:

- When there is a question about a possible mistake or imperfection of the writings
- To show the circumstances under which the instrument was made for the purposes of properly construing the instrument
- To explain an extrinsic ambiguity
- To explain the meaning of words existing within a written conveyance
- To explain conditions existing at the date of the deed

Another universal rule of construction to guide us in writing our descriptions is that in instances of discrepancies or variations in a document, the courts uphold spelled out words as opposed to abbreviations or numerical representations (such as "twenty" versus "20").

General rules of construction and of evidence in the United States have not changed much over the centuries from their British roots. Volume 1 of an 1846 treatise by George Spence, Esquire ("One of Her Majesty's Counsel") entitled *The Equitable Jurisdiction of the Court of Chancery*[13] provides a lengthy discussion of numerous rules of construction for deeds that still apply in more modern times. For example:

It is a rule that every man's grant shall be taken by construction of law most forcibly against himself....[14]

In furtherance of the rule that a deed shall have and take the most effect that may be according to reason, if the words are not properly arranged in order to give effect to the intention of the parties, the court in proper cases exercises the power of marshalling them so as to carry the intention into effect. ... Whenever the words of a deed, or of the parties without deed, may have a double intendment,

---

[13]The full title of this volume is *The Equitable Jurisdiction of the Court of Chancery; Comprising Its Rise, Progress, and Final Establishment; to Which Is Prefixed, with a View to the Elucidation of the Main Subject, a Concise Account of the Leading Doctrines of the Common Law and of the Course of Procedure in the Courts of Common Law in Regard to Civil Rights; with an Attempt to Trace Them to Their Sources; and in Which the Various Alterations Made by the Legislature Down to the Present Day Are Noted,* by George Spence, Esq., One of Her Majesty's Counsel, published by Lea and Blanchard, Philadelphia, 1846.
[14]Ibid., p. 527.

and the one standeth with law and right, and the other is wrongful and against law, the intendment that standeth with law shall be taken.[15]

Presumptions of law arising upon the words used may be rebutted by a plain intention appearing upon the instrument to the contrary.[16]

The rules of evidence are universally the same . . .; therefore, parol evidence offered for the purpose of substantially altering a written instrument, cannot be received in a Court of Equity any more than in a Court of Law.[17]

These words, written by a British subject in the middle of the nineteenth century, still serve to explain the operation and application of rules of both construction and of evidence today. American citizens have written similar treatises, and for our own view on the matter of construction and evidence we turn to *A Treatise on the Law of Real Property*,[18] written in 1896 by Leonard A. Jones, an American law scholar:

Where a general description is joined with a particular one, it is a rule of construction that the latter prevails over the former. A general description may be limited, restrained, or controlled by a particular description; but as a rule a particular description is not limited, restrained, or controlled by a general description. The real interest of the parties should, where possible, be gathered from the whole description. The calls in a deed, whether natural or artificial, are divided as regards their relative value into two classes, - descriptive or directory, and special locative calls. "The former, though consisting of rivers, lakes, and creeks, must yield to the special locative calls, for the reason that the latter, consisting of the particular objects upon the lines or corners of the land, are intended to indicate the precise boundary of the land, about which the locator and surveyor should be, and are presumed to be, very particular; while the former are called for without any care for exactness, and merely intended to point out or lead a person into the region or neighborhood of the tract surveyed, and hence not considered as entitled to much credit in locating the particular boundaries of the land when they come in conflict with special locative calls, and must give way to them.[19]

We can find innumerable cases repeating the axioms that law determines *what* a boundary is, while fact determines *where* that

---

[15]Ibid., p. 539.

[16]Ibid., p. 545.

[17]Ibid., p. 555.

[18]The full title of this text is *A Treatise on the Law of Real Property as Applied Between Vendor and Purchaser in Modern Conveyancing or Estates in Fee and Their Transfer by Deed,* by Leonard A. Jones, A.B., LL.B, published by The Bowen-Merrill Company, 1896.

[19]Ibid., Volume 1, §410, pp. 338–339, including a quotation from *Stafford v. King,* 30 Tex. 257 at 273.

boundary is. Similarly, when it comes to the documents describing those boundaries:

> Whether an instrument is ambiguous is a question of law. The construction to be given an unambiguous instrument is a question of law and not one of fact. These elementary rules are so well established as not to require the citation of authorities.[20]

There are two main approaches to construing deeds: the "four corners" approach and the "context" approach. We will examine them both. In either instance, the reader of a deed must be objective and seek the intent of the parties rather than imposing personal bias into the investigation.

6.6.2.2 "Four Corners Rule"   The rules of construction tell us to look first within the description for evidence of the intent of the written instrument of conveyance, and to read that document for what it says on its face. This process of looking within requires us to look at all parts of the document, all the way out to its four corners. We cannot arbitrarily ignore one part because it does not suit our purposes or choose one other part to uphold because it matches our preconceived ideas. The object is to harmonize all parts of the deed rather than dismiss any single part of the document from consideration. Returning to Spence's 1846 treatise on *The Equitable Jurisdiction of the Court of Chancery:*

> It is an old established rule, of universal application, that the construction must be upon the entire instrument, so that one part shall help to expound the other; and . . . so that *every word* (if it may be) may take effect and none be rejected.[21] . . . Further, the construction is to be such, that all the parts do agree together, and there be no discordance therein; and so that the whole instrument and every part may take effect, and as much effect as may be to that purpose for which it is made.[22]

Following the narrative descriptive part of a deed, we find a section called the *habendum clause*. This section begins with the words "To have and to hold" (the English translation of the Latin term *habendum,* meaning terms of holding or possession the land), and defines the extent of ownership to be held and enjoyed by the grantees. The habendum clause provides the basis for a grant of fee, and when it does not appear in a deed, the document is often considered to convey only a

---

[20]*Davis v. Andrews* (361 S.W.2d 419 at 424, Court of Civil Appeals of Texas, Dallas, 1962).
[21]Spence, supra, p. 527.
[22]*Ibid.,* p. 528.

lesser interest, such as an easement. However, if the phrase "and their heirs" is added in naming the grantees in the opening granting clause of the deed, the courts may determine that habendum clause can be completely omitted, as the phrase "and their heirs" accomplishes the same intent, tracing back to feudal roots of our modern real property conveyance system.

The language of the habendum clause is brief but to the point, the following being a common example preprinted on papers formatted for deeds that will be recorded in the County Clerk's office:

> To have and to hold, all and singular, the above mentioned premises, together with the appurtenances, unto the said party of the second part, its successors and assigns, to its own proper use, benefit and behoof[23] forever.

While not a section of the deed that surveyors write, the habendum clause may be important to explain, qualify, lessen, or enlarge the estate described and granted. If its language is found to be completely contradictory to the rest of the deed, the habendum clause is not upheld as having any greater weight than any other part of the deed in determining the intent of the document. It is, however, a section that is often turned to in arguing the extent of ownership granted or not granted in a deed.

The four corners rule tells us that the habendum clause cannot be taken out of context and must be read with the rest of the deed to form a whole picture of the intent. Again from Spence's treatise:

> If there be two clauses in *a deed* so totally repugnant to each other that they cannot stand together, the first shall be received and the latter rejected, for the first deed is always most available in law . . . In a deed, words in the *habendum* which are repugnant to the previous part, may be rejected.[24]

A good example of why the full four corners of the deed must be considered sets the stage for the case of *Davis v. Andrews* (361 S.W.2d 419, Court of Civil Appeals of Texas, Dallas, 1962). Davis was one of the heirs of the original 1930 grantors of a 50-acre tract of land in which the mineral interests had already been granted to F. L. Luckel, who in turn had issued a lease to the Pure Oil Company. Andrews was the successor in title to the grantee of the 1930 deed, which noted that sale was subject to the terms of the lease but that the grantees would receive a percentage of royalties, also according to that same lease.

---

[23] *Behoof* means benefit, advantage, or convenience, another reminder of our ancient English legal roots.
[24] Spence, supra, p. 536.

The habendum clause of the deed read as follows:

To have and to hold the above described property, together with all singular the rights and appurtenances thereto in any wise belonging unto the said Grantee herein, his heirs and assigns forever; and we do hereby bind ourselves, our heirs, executors and administrators to warrant and forever defend all and singular the said property unto the said Grantee herein, his heirs and assigns against every person whomsoever lawfully claiming or to claim the same or any part thereof, *for a period of 20 years from date hereof and no longer. [Emphasis supplied by the court.]*

Oil had been discovered on the site prior to 1930, and had produced continuously, with royalties paid to the landowners. But now Davis has claimed that the 20-year limit cited in the deed meant that the interests conveyed by the 1930 deed would terminate in 20 years after the deed transaction and revert (along with the related oil royalties) to the heir of the original grantor. Andrews replied that the limitation of 20 years applied only to the warranty expressed in the deed (the guarantee of title), and followed up by saying Davis and the other heirs were time barred by 8 years beyond the 20-year time frame Davis claimed.

The court identified various rules of construction that must apply to this case, the first and foremost being that it is not what the parties meant to say but failed to express, but instead the actual expression of intent in the deed that will control the interpretation of the deed. In other words, the question is not what the parties *meant* to say, but rather the meaning of what they *did* say that must be addressed. Unambiguous deeds are enforced as written, even though they do not express the original intent of the parties. Further, no one may pick and choose the isolated parts of a deed to uphold a personal preference in its interpretation; the document must be read and interpreted as a whole.

In addressing Davis's reliance on the habendum clause as a primary expression of the intent to limit the interests conveyed in 1930, the court said this:

Another and equally important rule of construction, sometimes called *"four corners rule"* is that the intention of the parties, and especially that of the grantor, is to be gathered from the instrument as a whole and not from isolated parts thereof.[25] *[Emphasis added]*

And, more specifically, regarding the means of construction:

A grammatical analysis of the habendum clause and warranty clause reveals that they are contained in a compound sentence, each complete in itself. It

---

[25]361 S.W.2d 419 at 423.

is a well recognized rule of grammatical construction that a qualified portion of a sentence qualifies that sentence only. The qualifying provisions, dealing with the limitation of 20 years, applies *[sic]* only to the warranty clause of the compound sentence. It necessarily qualifies only that part of the sentence relating to warranty and not as to grant.[26]

Certainly, this is a warning to take our English lessons seriously; the placement of a punctuation mark may play the starring role in a resolution of a description's meaning. And if the description does not appear ambiguous, then no extrinsic evidence, including parol testimony, will be allowed in attempts to alter that plain meaning. It is not what the grantors mean to say, but what they actually do say unambiguously that will control the interpretation of a deed or contract.

6.6.2.3 *Context*

One of the difficulties in construing descriptions is that we must interpret these documents according to conditions at the time they were written. This can require research into who owned what lands, locally used names for certain buildings or intersections, time-altered names of trees and geographic features, evolution of spelling and handwriting styles over the centuries, and a lot more detective work.

As a very early example of a deed on the western side of the Atlantic Ocean, the following document represents the culmination of William Penn's treaties with the Delaware Indians (also known as the Lenni Lenape) in what was to become part of Bucks County in southeastern Pennsylvania (or "Penn's woods"). William Penn fully acknowledged the claim of the Delawares to the land, a view very different from other European arrivals, and Penn attempted to deal with them fairly. This particular deed preceded the notoriously unfair Walking Purchase of 1737 (executed by Penn's heirs) that is far more familiar to readers of history.

23 June 1683

We, ESSEPENAIKE and SWANPISSE, this 23rd day of the 4th [sic] month called June in the year according to the English account 1683, for us and our heirs and assigns, do grant and dispose of all our lands lying betwixt Pennypack and Neshaminy creeks and all along upon Neshaminy Creek and backward of the same, and to run two days' journey with a horse up into the country as the said river does go, to WILLIAM PENN, proprietor and governor of the province of Pennsylvania, etc., his heirs and assigns forever, for the consideration of so much wampum and so many guns, shoes, stockings, looking glasses, blankets, and other goods, as he the said William Penn shall please to give unto us hereby for us, our heirs and assigns, renouncing all claims or demands of

---

[26] 361 S.W.2d 419 at 424.

anything in or for the premises for the future from him his heirs and assigns.
IN WITNESS whereof we have hereunto set our hands and seas the day and
year first above written.

Sealed and delivered in the presence of	Indians present
N[icholas] More	Weanappe
Lasse Cock	Enshockhuppo
Thomas Holme	Etpakherah
C[hristopher] Taylor	Alenoh
Thomas Wynne	The mark of X Essepenaike
	The mark of X Swanpisse[27]

So where exactly is this land, how much land is it, and is the description too vague to be considered legally sufficient? This document
definitely is in the form of an enforceable contract, naming the parties,
naming what is to be transferred, naming the consideration, and bearing the signatures (or marks) of grantors and witnesses. But can it be
located on the ground?

Definite monuments do appear in this deed. We know that the Native
Americans "owned" the land ("ownership" actually being a foreign
concept in a culture that more often occupied and stewarded the land
rather than claimed private rights in it), and there were reports in early
colonial records of where the Delawares lived, hunted, and farmed. This
deed references natural monuments, the Pennypack and Neshaminy
Creeks, which still run through the area. And the pace of travel by
horse over varying terrain was also known. While perhaps difficult for
retracers centuries later, the monuments in this deed provided all the
information necessary to locate the tract in 1683.

A few centuries later, we find an appeal from a U.S. District Court
decree in California that returned a private land claim to the public
domain in the case of *Higueras v. United States* (72 U.S. 827, Supreme
Court of the United States, 1864). Ignoring the complexities of the
legal process, the Supreme Court's decision underscores the importance
of monuments and of making readily apparent corrections to surveys
when the compass points are rotated from "true" directions. Please note
that it is unusual for the federal Supreme Court to become involved
in boundary disputes, but when the government considers the land
involved to be its own, it does take such matters to court, and it may also
consider cases when the problem is rooted in the division of lands under
government patents and the federal rectangular public land system.

---

[27] Jean R. Soderlund (Ed.), "Document 68, Deed from Delaware Indians." *William Penn and the Founding of Pennsylvania: A Documentary History,* Philadelphia: University of Pennsylvania Press, 1983, pp. 287–288.

The original claimant to the land (Jose Higueras) had acquired a possessory right to a tract of land called "Tularcitos" in 1821 by means of a decree from the governor of the territory. The decree ordered measurements (in varas) to be made and monuments fixed on the fours sides of the tract, and the assigned commissioner returned a record of his work to the government.

Because an adjoining landowner later claimed part of the tract, Higueras again petitioned the territorial governor again in 1835, requesting an enlargement of the boundaries of the original concession, and confirmation of his title to the first concession. The second request was based on Higueras's long occupation of the tract and on the fact that part of the tract in the decree had been granted to another person.

The governor granted this second decree of concession, annexing it to the first, along with the following new description of the entire tract, in conformance with colonization laws. The first decree appeared to grant no more than a possessory right, and the second was without formal title, but together, along with other government actions, gave Higueras a right to the tract.

> Description of the tract as given in the decree of confirmation is that it is situated in Santa Clara County and is the same land formerly occupied by Jose Higuera [sic], now deceased, and is known by the name of Los Tularcitos. Boundaries given in the decree are as follows: Beginning at the back side of the principal house on said rancho, standing at the foot of the hill, and running thence northwardly to a lone tree on the top of the sierra (which tree is known as a landmark), thence east along the sierra to the line of the land known as the rancho of Jose Maria Alviso, thence southerly along the west line of said Alviso's rancho till it intersects the Arroyo de la Penetencia, thence up said arroyo to an estuary, and from that point to the place of beginning.[28]

A great deal of our western past relating to the particular rules regarding land grants, decrees, and patents plays out in this particular case, and related "quality of title" issues are omitted here for the sake of brevity. However, during this phase of American history, it was the duty of the surveyor-general to survey and plat all private land claims that were confirmed by decree. District Courts were also authorized to order examination and adjudication of any survey of a private land claim when any interested party (like Higueras) applied. The surveyor-general completed his survey and returned it to the District Court in June 1859, but Higueras disagreed with this survey.

---

[28]72 U.S. 827 at 830.

The District Court confirmed the survey after a hearing, and the present suit appealed that confirmation. The United States was a party in the suit because invalid claims return the land to the public domain,

In evaluating the boundaries of Higueras's decree of confirmation, the court noted that "Most or all of the courses given are undoubtedly erroneous, but there is little difficulty in ascertaining the cause of the error. . . ."[29] Bearings on two maps of the tract were rotated so that north on the map was actually northwest, but simply making this rotational correction placed the monuments referenced in the decree in agreement with the courses on the maps. The Mexican witnesses in the case (remember that California was formerly part of Mexico) had made similar directional errors in their testimony, but they knew the monuments designated as marking the boundaries of Higueras's land. Similarity between the language of the decree and of the witnesses' testimony made it evident that the commissioners had relied on local identification in writing the description for Higueras's boundaries.

No one disputed the point of beginning or the first course of the description. The monuments called for in the decree were well known and proven, and Higueras could not claim that they were intended to be anything other than corner boundaries. The intention of the decree was obvious. In analyzing the evidence and calls in each line of the decree, disregarding the error in the compass direction (easily reconciled by rotating the entire basis of bearings), the Supreme Court found the calls complete and specific, and that the survey corresponded with the decree.

And so we find that local knowledge provides a contextual backdrop against which we can evaluate the consistency of the written record and notoriety of landmarks reported on surveys. The local perception of the direction of "north" was the key to harmonizing physical and record monuments with the written description.

Context to understand a description may be found in local use of language, local names for trees, and local practices. We find historical backdrops in contemporaneous legislation, such as the process of confirming decrees in the *Higueras* case. The kind of equipment in use at the time of the survey may affect measurements of angles and distances; certainly the levels of both accuracy and precision varied from current standards, but we must not project our modern expectations backwards through time to judge the quality of our predecessors' work. If the work met or exceeded standards in place when the survey

---

[29]72 U.S. 827 at 834.

was conducted and the description is clear and complete, we uphold that work.

In parts of the country, the process of surveying is subject to local standards and procedures, much more specific in scope than minimum surveying standards established by state regulatory boards or state statutes. In Pennsylvania, the city of Philadelphia established the first positions of surveyors and regulators by an Act of March 26, 1762,[30] and expanded these posts throughout the city over the following decades. In 1854, the offices of district surveyors and regulators were moved to the city's Department of Streets. The newly revamped positions were first filled by election (both the number and the method of filling the posts have changed over the years), each of twelve originally designated districts naming as its Surveyor and Regulator a "citizen, who shall have had five years' experience and skill in his profession,"[31] all such District Surveyors together forming the Board of Surveyors with a thirteenth member being the Chief Engineer and Surveyor (in the days before either profession was licensed separately).

The Philadelphia Board of Surveyors has had a role in city planning since its early years, including the authority to alter lines and grades of public streets and decide all questions regarding party walls (position and thickness). All "public plans of town plots"[32] were moved to the office of the Board of Surveyors early on, and this office is still the repository for much information ranging from surveys of land incorporated into the city to elevation data to the length of 100 feet. While this last may sound peculiar to those unfamiliar with the practice, there is often a difference between 100 U.S. feet and 100 district standard feet, and the Philadelphia District Surveyor and Regulator disseminates this information. If reading a deed for a tract within the city's limits, one had better be aware of the fact that the reported measurements in the description are often longer on the ground, and not consistently so from one district to another. In the Fifth District, 100 feet will measure 100.25 in U.S. standard feet, while in the Eighth District 100 feet district standard is equal to 100 feet $2^1/_2$ inches (*not* reported as 100.06 feet) in U.S. standard.

---

[30]*The City Government of Philadelphia: A Study in Municipal Administration,* prepared and published by Wharton School of Finance and Economy, University of Pennsylvania, Philadelphia, 1893, p. 139.

[31]*A Digest of Laws Relating to the City of Philadelphia, from the Territorial Extension, by Act of Assembly, Approved February 2d, 1854, Until the Close of the Session of the Legislature in 1855,* prepared "under the supervision of the Law Department of the City of Philadelphia," published by King & Baird, Printers, 607 Sansom Street, Philadelphia, 1865, p. 48.

[32]Ibid., p. 49.

The federal territory known as the District of Columbia also has a history of district surveyor and regulator positions, and it is only since 1991 that private surveyors have been able to fully practice within the 10-mile square District. Among other cautions, the *Manual of Practices for Real Property Surveying in the District of Columbia*[33] notes that in parts of the city, recorded dimensions will usually differ from measured distances, and that sides of squares will not have parallel sides despite what is shown by subdivision records and plats. Even District ordinances outline how to address deficiency or excess in measuring squares in the City of Georgetown within the federal District.[34] Obviously, one cannot walk into this situation and expect to follow a description easily. The local context is imperative to understanding differences between written record and what is found on the ground.

Outside of the original 13 colonies, the government agency regulating the public lands of the United States issues instructions on how to conduct most of the variations of the rectangular system of surveys. While the Land Ordinance of 1785 specified how lands in the "Western Territory" were to be disposed of, the first surveys were not guided by written directions, as later work would be. We can, however, find first letters providing instructions (soon after the first Land Ordinance surveys) and later "manuals of instructions" as the rectangular system evolved and expanded westward across the continent. The instructions in effect at the time of the original survey must guide later retracers of government lines (although these are not necessarily property lines).

Because of the correlation between private patents and the public land survey system, the historical context of instrumentation, methodology, and government dictates provides an important background when deciphering surveys and descriptions originating at different times—and even in different parts of the country. Early rectangular surveys in Ohio and Kentucky served as testing grounds for finding a reasonable approach to dividing lands. Conveyances from the Ohio Company (beginning in 1786) and those subject to Military Reserve Surveys provide just two examples of variations from the current 36-section system that features a consistent serpentine numbering system and establishment of aliquot parts.

Documents and treaties regulating the Louisiana Purchase included specific requirements for surveys of private estates established under

---

[33]Published and distributed by the government of the District of Columbia, Department of Public Works, Design, Engineering, and Construction Administration, Office of the Surveyor, 1989.
[34]District of Columbia Code §1-1324.

Spanish and French monarchies. Land owners in the *Isle of Orleáns,* a specific region of southeastern Louisiana, had their private claims surveyed and verified by the Government Land Office. It is not uncommon for numbered sections in these townships to extend into the hundreds.

6.7   DEED DISCREPANCIES—CONFLICTS

While the hierarchy of calls provides guidance in establishing the references we should include when writing descriptions, it is not unusual to find a discrepancy between what a deed says and what is found on the ground. In such instances, returning to the intent of the grantor is made more difficult, but not always impossible. Numerous cases heard in numerous courts across the nation have upheld certain principles that can assist us in writing good descriptions and in interpreting existing recorded documents.

In the previously discussed case of *Higueras v. U.S.*, the United States Supreme Court observed:

> But ordinarily surveys are so loosely made, and so liable to be inaccurate, especially when made in rough or uneven land or forests, that the courses and distances given in the instrument are regarded as more or less uncertain, and always give place, in questions of doubt or discrepancy, to known monuments and boundaries referred to as identifying the land. Such monuments may be either natural or artificial objects, such as rivers, streams, springs, stakes, marked trees, fences, or buildings.[35]

The "looseness" of surveys to which the court refers is closely related to the vast undertaking of surveying patents in newly acquired territories and the entire public lands. Inhospitality of land, weather, and Native American residents could make the process difficult, highly subject to error, and, although not mentioned by this court, possibly fraud. The difficulty of making measurements under arduous conditions with battered equipment lent even more weight to known, identifiable, and referenced monuments as evidence of intended boundaries. We are sent back to the hierarchy of calls rather than settling for the mathematics and measurements of directions and distances.

Ayers v. Watson is another of the infrequent boundary dispute cases heard by the Supreme Court of the United States. Ayers, the original defendant, lost six jury trials and a circuit court trial, then went on to

---

[35]*72 U.S. 827 at 835–836.*

lose three United States Supreme Court trials as well, but his attempts elicited some clear directives from the Supreme Court about the significance of original intent, the hierarchy of calls, and how to approach discrepancies when deed calls do not match deed courses and distances.

Ayers claimed that his title to a tract of eleven square leagues of land in Texas was older and superior to Watson's title. Ayers based his claim on a grant by the government of Coahuila and Texas to Maximo Moreno in 1833. Watson, who had filed to eject Ayers from the disputed area, claimed land with its title rooted in a patent from the State of Texas given to Daws. Ayers counterclaimed that Watson's claim was merely "some pretended patent" that overlapped his own claim by a third of a league.

The first arguments before the United States Supreme Court (113 U.S. 594,1885) revolved around the admissibility of certain evidence and instructions given to the jury during the trial court hearing. The primary issue was the identity of two hackberry trees called out in Watson's deed: were the two trees found in 1854 along the eastern line of the tract the same hackberry trees mentioned in Watson's deed? The trees found in 1854 were not exactly at the point arrived at if the courses and distances of Watson's deed were followed strictly.

Watson's deed extended back (northwardly) from the river 12 to 14 miles, and the question was whether this distance overlapped Ayers's claim. If it did, then Ayers would prevail because he held an older (senior) title. If there were no overlap, then Watson would be entitled to the land. The burden of proof was on Ayers to show that his 11-league tract extended back from the river far enough to include all or any part of Watson's land.

There was no dispute about beginning point of the 1833 Moreno grant to which Ayers traced his title. The court noted, however, that following

> ... the lines of the survey by courses and distances only, it would embrace nearly the whole of the Daws patent; but, run in this way, the lines would not coincide with certain well ascertained monuments, either called for in the grant, or conceded to mark and identify the footsteps of the surveyor who originally located it in 1833.[36]

These well-known features included several miles of marked trees and the river San Andres.

---

[36] 113 U.S. 594 at 601.

The jury had been instructed to "follow the tracks of the surveyor," as well as could be reasonably ascertained, and only to follow course and distance when the marks made by the surveyor in 1833 and the natural monuments the surveyor referenced in his field notes could not be found. Specifically, marked trees would control both course and distance when those marked trees had been called for in the survey and the deed prepared based upon that survey. However, Ayers had wanted the trial court to instruct the jury that "following the footsteps" meant adhering to the measurements identified in the 1833 Moreno survey, rather than honoring references to natural or artificial monuments.

The court declined to do so, stating:

It has been repeatedly held by the Supreme Court of Texas, *as a general rule*, that *natural objects called for in a grant*, such as mountains, lakes, rivers, creeks, rocks, and the like, *control artificial objects*, such as marked lines, trees, stakes, etc., *and that the latter control courses and distances*. . . . There are exceptional cases, however, in which courses and distances may control, as where mistakes have been made by the surveyor as to objects called for, or where the calls for monuments are inconsistent with each other and cannot be reconciled, or where some other clearly sufficient reason exists for disregarding the general rule.[37] *[Emphasis added]*

Clearly, the court recognized that not every survey is perfect (although giving surveyors the benefit of the doubt in calling those "exceptional cases"), and in such instances the general rules will yield to the weight of evidence for a different approach to interpreting a description. This quotation also underscores the need for clear identification in a description as to the intended objects or monuments that are meant as markers of a tract's boundaries. A common modern dilemma is created by the use of unexplained abbreviations to describe markers at the corners of a surveyed tract. "IP" is one such problem. Does this mean "iron pipe," "iron pin," "iron post," or "irreconcilable problem"? Beyond straightforward identification of the object, its dimensions and markings should be included to allow the next reader of the deed and survey to trust that the marks found in the field are the same as mentioned in the deed. It is, after all, a surveyor's responsibility to place trust only in original, readily identifiable, and undisturbed monumentation. A complete and well-written description furthers that process.

Ayers was unhappy with his loss, and returned for a second hearing by the Supreme Court (132 U.S. 10, 1889). This time the court addressed the validity of accepting testimony of deceased witnesses, as

---

[37] 113 U.S. 594 at 605.

Ayers's surveyor had died since the first case and Ayers wished to rely on his former surveyor's deposition from an earlier and unrelated case about the same tract of land. The court barred that evidence as the late surveyor would not be able to explain his prior testimony, which might have had an error in it, and he could not be cross-examined. This was only an issue because of discrepancies between the surveyor's 1860 deposition and his more recent ones in 1877 and 1880.

But also in this opinion, the Court upheld the trial court's instructions to the jury (protested by Ayers) that the point of beginning carries no more weight than any other corner in a survey, and that courses may be reversed and run backwards if that will help to harmonize calls in the grant and to meet the intent of the grant.

In Ayers's third appeal (137 U.S. 584, 1891), the Supreme Court once again stated that:

> . . . the footsteps of the original surveyor might be traced backward as well as forward; and that any ascertained monument in the survey might be adopted as a starting point for its recovery. This is always true where the whole survey has been actually run and measured, and ascertained monuments are referred to in it.[38]

Clearly, monumentation that is set as part of the original survey and then referenced and fully described in the deed between the original parties best identifies the intent of those parties.

---

[38] 137 U.S. 584 at 590.

CHAPTER 7

ALTA/ACSM SURVEYS

7.1 LAND TITLE INSURANCE

In anything other than a quitclaim transaction, contracts for the sale of real property require that the seller convey marketable title to the purchaser. In defending title, any defects in the title of record as well as hidden defects not disclosed by public records must be considered, along with the cost of defending the title against attack by other claimants. In the United States, title insurance evolved in response to inconsistent land recordation processes and laws as a means to insure against financial losses resulting both from defects in title and unenforceability or invalidity of mortgage liens against title.

In section 1.4 we discussed the different statutory approaches to recording title documents, each with its own benefits and drawbacks. In searching titles to real property interests we find that not all transactions are recorded, or they may be recorded and indexed incorrectly so that we have difficulty finding them, or they may be recorded properly but contain errors in their contents. Most difficult to discern are the "hidden defects," which include but are not limited to the following[1]:

- The legal disability of a grantor in a chain of title
- Forgery of a deed, mortgage, or other instrument in the chain of title

---

[1] John E. Cribbett, William F. Fritz, and Corwin W. Johnson, *Cases and Materials on Property*, 2nd ed. Brooklyn: Foundation Press, Inc., 1966, p. 872.

- Fraudulent representation of marital status by a grantor in the chain of title
- Mistaken identity of a record title holder and a grantor due to similar or identical names
- Errors in the record
- Errors in the examination of the record
- Undisclosed heirs
- Exercise of a power of attorney after death of the creator of the power
- Defects in conveyances in the chain of title due to a lack of delivery to the grantee

In addition to these hidden defects we can add the differences of opinion regarding interpretation of a deed as another source of litigation and expense.

Title insurance covers damages only for the risks that the insurer knows about, and policies eliminate risks by including requirements that must be fulfilled prior to issuing a policy and listing exclusions from coverage. The records searched by the insurer are only those of public record, and do not always reveal physical conditions of the property that may impact the marketability of the property. For this reason, title policies often include endorsements accepting responsibility for and making them subject to encroachments or adverse circumstances and conditions affecting title that would have been "disclosed by an accurate survey" (or similar language, depending on the insurance company). Those later disclosures are exceptions from title insurance policy coverage unless coverage is added with the endorsement.

7.1.1 Why a Survey Matters

A case frequently cited when discussing survey exceptions from title policies is *Walker Rogge, Inc. v. Chelsea Title & Guaranty Company* (562 A.2d 208, Supreme Court of New Jersey, 1989). Walker Rogge, Inc. had contracted to buy a tract of land based on a "per acre" cost, expecting it to contain "19 acres more or less" as per a 1975 survey. After requesting an updated survey, closing proceeded in December of 1979, with the grantors using the 1979 survey description of 18.33 acres rather than 12.43 acres as per the deed by which they had acquired title.

In January 1980, Chelsea Title & Guaranty (Chelsea hereafter) issued a title insurance policy containing a standard language exception against loss or damage by reason of "encroachments, overlaps, boundary line disputes and other matters which could be disclosed by an accurate survey and inspection of the premises." Usually, this exception to insurance coverage is removed from the title policy when the title company is presented with a current survey.

It was not until 1985, when Walker Rogge became interested in purchasing the adjoining lot to combine with its original purchase that it discovered a problem with the survey of its first purchase. On hiring a different surveyor to perform this second survey, the error in the area of the 1979 survey became apparent, and Walker Rogge sued upon realizing it had paid for nearly six acres it had not acquired.

While not going into the details and outcome of this suit, the one before it, and the one on remand after it, the single point we wish to make here is that there is a difference between examining records and a survey on the ground. There is also a difference between "an inspection of the premises" and a survey on the ground. These differences are why a survey is so important in the real estate transaction process. We quote from the court's opinion:

A survey and inspection serve related but different purposes. The purpose of an inspection, which is performed by merely visiting the property, is to disclose such matters as physical encroachments, evidence of adverse use, and monuments. By comparison, a survey, as this case demonstrates, can involve extensive research and field work. Unlike a mere inspection, a survey relates the property as described in recorded instruments to the land as it exists. ... Furthermore, an inspection can be performed simply by visiting the property. It would distort the title policy to the point of illogic to expose [the title insurer] to the risk of the results of an accurate survey merely because those results could not be revealed by an inspection of the premises.[2]

7.2 ALTA/ACSM SURVEY STANDARDS

To ensure that it is relying on "an accurate survey," the title industry has joined with the national professional surveyors' organization to create a set of standards for preparing surveys and checklists for informing those who order surveys of the variety of conditions that may be encountered. The American Land Title Association (ALTA) is a national organization of title abstractors and title insurers. The American Congress on Surveying and Mapping (ACSM) and its surveying member organization National Society of Professional Surveyors (NSPS)

---

[2]562 A.2d 206 at 217.

have collaborated for many years with ALTA to create a set of standards to guide surveyors in meeting the needs of title examiners and insurers. The purpose of these joint standards is to provide the "accurate survey" on which the title insurer can rely; there is no insurance coverage against mistakes made by the surveyor, and the policy addresses only the factual revelations made by the survey.

The document containing the standards is entitled "Minimum Standard Detail Requirements for ALTA/ACSM Land Title Surveys" and is available through the Internet websites of both organizations. It is reviewed periodically to conform to current practices in both the title and surveying professions and to address possible difficulties or confusions arising from prior versions. The surveys conforming to this set of "minimum standard detail requirements" are referred to as "ALTA/ACSM Land Title Surveys" or more briefly as "ALTA surveys."

Because of the level of detail involved in the required investigation and reporting for ALTA surveys, the cost for performing such work is generally greater than for "ordinary" real property surveys. For this reason, it is most often commercial sites and high-priced residential properties for which ALTA surveys are requested.

In general, the ALTA/ACSM standards establish a means of contracting for a land boundary survey that will meet the needs of all parties to the real property transaction. The concept of "contract" is central to this process, as this is a highly specialized request that both the requestor and the surveyor must understand and agree to regarding contents and presentation. The document containing the full "minimum standard detail requirements" clearly identifies what is mandatory and what is optional regarding what the request for a survey is to include.

7.3 MANDATORY REQUIREMENTS FOR ALTA SURVEYS

ALTA standards are meant to accomplish two things: (1) guide the surveyor in supplying that information in a complete and consistent manner; and (2) guide the requestor of an ALTA survey in securing the kind of survey needed for a specific site. The first part of the specifications generally address the standards expected of both requestor and surveyor, and Table A is provided as a means to guide the requestor and surveyor through the negotiations for contracting an ALTA survey.

At the time of this writing, the 2011 version of the document is in effect. If you are performing or requesting ALTA surveys, you must acquire the most recent version of the document and *read it*. This is necessary for the protection of both parties involved in the contracting process. Both the contents and the definition of terms have changed over time, as evident in the now-defunct land use classifications as the defining factor in establishing positional relationships.

The ALTA standards are set up first as numbered paragraphs with mandatory requirements of the surveyor and the one requesting the surveyor (the "client") and then as a table of optional additional services that the client may request if not already part of the state-regulated practice of surveying in the state where the property is located.

The first mandatory requirement has not changed over the years beyond the tweaking of language. It requires a *written* contract between the client and the surveyor, which (although not stated) makes the contract enforceable against both parties. The client shall supply either the record description of the property to be surveyed or of the parent parcel containing the site if this is to be an original survey creating a new tract. The client is responsible for providing all record documents affecting the site, including documents of record referenced in other record documents. The surveyor is to examine all these documents and note them on the plat or map of survey.

The next several mandatory requirements are for the purpose of identifying who prepared the survey, identifying the standards adhered to in performing the survey, and assuring that the survey plat or map is legible and understandable. These requirements address graphic scales, orientation of the drawing, and explanation of symbols and abbreviations.

The most comprehensive set of these requirements first establishes that the ALTA survey is in fact a survey made on the ground, rather than a compilation of record data reported in graphic form. Angles, bearings, and distances both measured and of record are reported (noting their sources) and compared. Physical markings, natural features, evidence of use or possession: all must be reported. As of this writing, any visible improvements within five feet of each side of boundary lines are to be noted on the survey, as these may in fact be physical evidence of encroaching or adverse use affecting the title to land.

While the surveyor and client may contract for something more, ALTA standards establish the minimum requirements for what is to be delivered to the client upon the completion of the survey work. As part of the package of deliverables, the surveyor must provide a report, either in the form of notes on the survey document or as a separate attachment, addressing differences between survey results and the title record, describing the analysis of evidence, and explaining any failure to meet relative positional accuracy requirements.

The standards acknowledge that riparian boundaries are volatile. Surveyors are required only to document when and how the location of a water boundary is shown on the map or plat, although the extent of any known changes shall also be identified. The surveyor is generally not required to research past locations of these changeable lines.

Finally, the mandatory requirements include the form of the certification that is *required* to appear on ALTA/ACSM Land Title Surveys. Often, state or local regulatory agencies dictate a specific certification as well. In such instances, the surveyor typically will place both certifications above his signature. Certifications that are not required by the ALTA contract or governmental regulatory authorities *may not be used*. The wise surveyor will not include such catch phrases as "true and correct" or "any and all" in his certification. Professional liability insurers are a good source of information regarding appropriate language that does not extend the surveyor's liability.

7.4 ACCURACY STANDARDS

The idea of "what an accurate survey would disclose" may not include every possible threat to title, as in the instance of disputed ownership of land. Instead, accuracy standards relate to the measurements made and reported by the surveyor. Early ALTA survey standards classified the required degree of precision and accuracy according to the intended use of the land as follows[3]:

> *Urban Surveys* - Surveys of land lying within or adjoining a city or town, and including commercial and industrial properties, condominiums, townhouses, apartments, and other multiunit developments, regardless of geographic location
>
> *Suburban Surveys* - Surveys of land lying outside urban areas and developed for single family residential use
>
> *Rural Surveys* - Surveys of land such as farms and other undeveloped land outside urban and suburban areas which may have a potential for future development
>
> *Mountain and Marshland Surveys* - Surveys of land normally lying in remote areas with difficult terrain and normally having a limited potential for development.

---

[3]"Minimum Standard Detail Requirements for ALTA/ACSM Land Title Surveys as Adopted by American Land Title Association, American Congress on Surveying & Mapping, and National Society of Professional Surveyors," 1997.

Each of these classifications of land use denoted a relative value of the land, and the minimum angle, distance, and survey traverse closure requirements for survey measurements reflected that relative value. The "positional uncertainty" and "positional tolerance" cited for these classes of surveys were increasingly relaxed as the land use dropped from urban to suburban to rural to mountain/marshland survey. The standards defined the relevant terms as follows[4]:

"Positional Uncertainty" is the uncertainty in location, due to random errors in measurement, of any physical point on a property survey, based on the 95% confidence level.

"Positional Tolerance" for a specified Class of Survey is the maximum acceptable amount of Positional Uncertainty for any physical point on a property survey relative to any other physical point on the survey, including lead-in courses.

Over time, as land has become a scarcer commodity and measurement technology has improved dramatically, the classes of land use have disappeared from the ALTA standards and "positional uncertainty" and "positional tolerance" have evolved into "relative positional accuracy," defined as follows[5]:

"Relative Positional Accuracy" means the value expressed in feet or meters that represents the uncertainty due to random errors in measurements in the location of any point on a survey relative to any other point on the same survey at the 95 percent confidence level.

This standard is a statement of the precision of the survey procedures, not a certification of the accuracy of the results. The surveyor performing a modern ALTA/ACSM Land Title survey will be well advised to become fully proficient in spatial data analysis.[6]

Prior to the acceptance of data acquisition analysis as the proper method to evaluate survey information the industry depended upon "closed traverse" theory. It was thought that if a vertical or horizontal traverse "closed", then the positional data had to be reliable. A more correct interpretation of a closed traverse is this: It only *likely* a traverse that "closes" does not contain any gross blunders. The value derived from closure computations is simply the arithmetic sum of all the errors.

[4]Ibid.
[5]"2005 Minimum Standard Detail Requirements for ALTA/ACSM Land Title Surveys as Adopted by American Land Title Association and National Society of Professional Surveyors."
[6]For further reading about spatial data analysis, we recommend Charles D. Ghilani's 2010 text, *Adjustment Computations: Spatial Data Analysis*, 5th ed. Hoboken, NJ: John Wiley & Sons.

It is not correct to believe that a closure of 0.0 is without error or that a closure of 0.01 feet proves that every location computed in the traverse will be within 0.01 of the "true" value. Consider the following equation:

$$A + B - C - D = 0.01 \text{ foot}$$

What is the value of B? If I were to stipulate that the "true" value of the sum of A, B, C and D is zero, are you any closer to discerning the value of B? The closure computed for a traverse, vertical or horizontal, is the sum of multiple and diverse errors. Random errors *tend* to cancel each other out, systematic errors *tend* to be accumulative and blunders *tend* to produce large discrepancies. Random errors are not guaranteed to be offsetting. Systematic errors are not always accumulative, particularly when more than one type of systematic error is present. Blunders can be large, small, accumulative or, worst of all cases, compensating.

Let us assume that we "knew" the exact values (errors) in the equation above. If A = 0.01′, B = –0.01′, C = 0.01′ and D = –0.02′ the sum of the errors would be an indication of the relative accuracy of the individual measurements. However, if A= 0.21′, B = –0.21′, C = 0.01′, and D = –0.02′ the sum of the errors is the same (0.01 foot), but the relative accuracy between A and B is far different from the sum of the errors, representing a variation of 0.42 foot that cancels out in mere mathematical analysis and may not be noticed if individual figures are not examined carefully. This example can claim to express "accuracy" because we "know" the values presented in the example.

The users of survey plats often demand an evaluation of the "accuracy" of the values on the plat. Unfortunately, accuracy can only be determined if one has knowledge of the "true" value. We can never know the truth, so the absolute accuracy of any measured value is unknowable. We can control the procedures used to collect data and we can statically evaluate the *probable* results of those procedure. The concept of "relative positional accuracy" (more correctly relative positional precision, or RPP) has been adopted by the ALTA/ACSM 2011 Standards as the means of evaluating the *quality* of the data presented on a survey plat. Relative positional accuracy (RPA) or RPP is the value expressed in feet or meters that represents the uncertainty due to random[7] errors in measurements in the location of any point on a

---

[7]Positional uncertainty is the result of all errors, not just random. The theory of statistical confidence computes the RPA under the assumption that all sources of systematic errors have been identified and accounted for, and that there are sufficient procedural safeguards against blunders.

survey relative to any other point on the same survey at the 95 percent confidence level.

7.5 INFORMATIONAL OPTIONS

The process of negotiating a contract for an ALTA survey is not restricted to the mandatory requirements. The client may want additional information, and the standards provide a checklist of specific details that shall be included in or, by omission, excluded from the survey. Each additional item checked for inclusion increases the effort required of the surveyor and the cost of the work. Some of the items require the client to provide additional information or reference to external sources such as regulatory authorities.

While some of these items are optional as far as the ALTA/ACSM contract is concerned, they may be mandatory under state and local laws, rules or regulations. Often, persons ordering an ALTA survey do not fully understand the optional services listed or the significant costs associated with some of them. It is vital that the surveyor review the items requested and their costs with the party responsible for paying the bill.

7.6 THE DESCRIPTION FOR AN ALTA/ACSM SURVEY

As part of the mandatory deliverables for an ALTA survey, a description must appear on the face of the plat or map of survey or otherwise accompany it. This can be a repetition of the record title description, a description provided by the client, or a new description prepared by the surveyor. The complete and detailed metes and bounds description should be provided on legal-sized paper as an Attachment "A" and recorded with the deed documents. Frequently, survey plats either are not recorded with the deed documents or are reduced to letter or legal size and rendered illegible. Detailed metes and bounds descriptions that are placed on the face of a survey plat add no information to the plat, encourage excessive brevity, and are lost when the plat becomes separated from the deed documents or is reduced to legal size.

In the case of United States Public Lands and their aliquot division only, the description on the face of the plat must contain the aliquot division(s), Section number(s), Township and Range number(s), the controlling meridian, land district (if applicable), county, and state. For example:

> SE $^1/_2$ of Section 8, T3N, R5W, Louisiana Meridian, Northwestern Land District, St. Mary Parish, Louisiana.

In the case of a platted subdivision the description must contain the lot designation, block (if any), subdivision name, recordation data, city (if applicable), county, and state. For example:

> Lot 22, Block A, Sunshine Estates, recorded June 6, 1996, Conveyance Office Book 23, Folio 345, Jefferson City, Washington County, Missouri.

In the case of a metes and bounds land record system description must include the name of the present owner, recordation data, city (if applicable) county and state. For example:

> Wilber Franks Tract, recorded May 3, 1923, Original 54, Bundle 1928, Hamilton County, New York.

7.7   THE SURVEYOR IS IN CHARGE

Surveyors performing ALTA surveys must adhere to state regulations and standards as well as the ALTA/ACSM requirements. State and local laws cannot be overcome by the contractual requirements of an ALTA survey, and the professional undertaking such an assignment must be sure to uphold state-imposed responsibilities. This may require some alterations to the ALTA contract. When ALTA and governmental regulations conflict, the outcome shall be settled in favor of the more restrictive requirement.

A professional surveyor may not contract away performances required under governmental regulatory authority. For instance, in some states the licensed professional surveyor is responsible for the record research on which the survey will be based. Surveyors bound by such regulations may not rely exclusively on the title commitment documents that list easements and restrictions without finding themselves out of conformance with the terms of their licensure. In other instances, state regulations require surveyors to set monumentation at any property corners missing such markers when the survey work is for the purpose of conveying real property interests. Whether or not the requestor of an ALTA survey has asked for physical property markers, these surveyors must set them.

Most real property boundary surveys are performed for the purpose of transferring real property rights. Title attorneys and insurance companies are charged with the responsibility of providing a clear and unencumbered transfer of the real property rights associated with the subject tract. Title attorneys typically want the distances and directions reported on a modern survey to match exactly with the figures presented in the historical deed or old survey. These professionals are not qualified to make decisions relating to the interpretation of physical evidence of property locations. The surveyor must resist efforts by others involved in the title transfer to influence decisions pertaining to the location of corners.

The ALTA contract also specifies the form and wording of the certification that the surveyor must attach to the plat. This certification must be the only certification present and worded exactly as stipulated in the ALTA contract, with the singular exception of such additions as may be required by local governing authorities. Title attorneys and insurers are not to add or amend the wording of the certification in any manner; nonsurveyors often urge the use of words or phrases such as "true and correct" or "any and all" as a means of shifting the responsibility away from themselves. Even in cases of certifications that are not part of an ALTA contract, the surveyor is the sole authority on the wording of that certification. A certification should be treated with the same respect and consideration as any sworn affidavit.

CHAPTER 8

SITUATIONAL AWARENESS

8.1 DEED DISCREPANCIES—CONFLICTS

While we have discussed "rules of construction" when there is a discrepancy between the deed and what we find on the ground, one of the primary professional responsibilities of the surveyor is to read the documents as objectively as possible. This means that no matter how badly we would like to honor certain monumentation we find in the field that is near the presumed boundary lines and corners of the tract, we cannot impose our personal preconceptions or "I know better" attitude into the process of interpreting a deed in light of the intent of the parties.

When there are conflicts between elements within a given deed, or between deeds on opposite sides of a disputed lot line, the courts return to the hierarchy of evidence time and time again. For this reason, a complete description that includes evidence found both in the field and in the written record assures that the description's introduction into a courtroom will present the fullest opportunity to defend real property interests. If it does not mention monuments, if it fails to reference recorded subdivisions or prior conveyances, if it falls short of clearly stating the specific intent of the parties in words, the court need not listen to presumptions about what the description meant to say but did not express, and proffered extrinsic evidence can be objected to and excluded from the arguments.

8.2 PROFESSIONAL RESPONSIBILITIES

The general duties of the surveyor require the professional to:

- Gather evidence and report the facts
- Follow the footsteps of the original surveyor, not the last retracer or the steps that "everyone else" follows or the steps most easily followed
- When authorized by statute, courts, or party consent, locate new boundary lines (partition, boundary line commission, subdivision)
- Monument new boundaries when new land subdivisions are created (preserve and perpetuate the evidence)
- Provide appropriate documents for client purposes (conveyance, subdivision, consolidation)
- Educate the client through thorough survey report (plan and language)
- Identify location, geometric configuration, area

Part of the protection and preservation of boundaries involves a written record supporting the facts on the ground. Remember, where a boundary is and what a boundary is are two different questions, the first being a factual matter for the surveyor—and the jury—and the second being a legal matter to be determined by the judicial system. While the purpose of a lawsuit may be to settle a disputed boundary, the court's decision is not always preserved for future landowners to rely upon. Always be sure that the legal team requests the judge to order a new description be prepared and recorded to preserve the decision and prevent a repeat of the litigation just completed.

8.2.1 Understanding Historical Context

Writers and certain readers (including surveyors and title searchers) of descriptions both have a professional responsibility to understand the context surrounding the creation of a particular parcel. Presumptions of uniform standards and methodology in surveying the described tracts mean that preconceptions dictate the outcome—not at all the proper effort to follow the original surveyor's footsteps or the intentions of the original grantor and grantee. Sometimes a little historical research opens a new avenue of understanding. This is particularly so in areas originally surveyed during colonial times, as each imperial colonizer brought its own form of instruments, its own methodology of utilizing them, and its own units of measurement. In areas where land values have increased over time, sometimes dramatically so, the differences between modern and original methods of surveying and describing lands can add to or be the primary source of much frustration.

As just one relatively recent example, St. John in the American Virgin Islands had long been considered of insignificant value, despite its early possession by Denmark in the late 1600s and a few skirmishes between England and Denmark over ownership in the following century. The thin soils were hard to farm, despite the Danish diligence in establishing plantations, and upon the abolition of slavery to supply the continuous hard labor necessary to make these undertakings succeed, the plantations failed, replaced by small scale subsistence farming. The United States bought St. John from Denmark in 1917, but the present National Park status of thousands of acres did not occur until a donation by Laurance Rockefeller in 1956, followed by increasing tourism and eventually growing land speculation in this formerly overlooked island and its underwater beauty.

Toward the end of the 20th century, economic pressure finally exploded into a bitter conflict between two long-settled island families, the Georges and the Sewers, and a real estate development company from the mainland, Newfound Management Corporation.[1] The East End Quarter of St. John where the Sewers and Georges had owned land for over a century had long been considered remote, but development pressures brought this suit to a head.

The two island families had settled a boundary dispute with the aid of I. Anderson, a Danish surveyor, in 1893, who made no measurements but merely provided an agreed boundary with compass bearings, marking the agreed line with stone piles. At the time, such methods were common practice in dividing low-value lands. The surveyor's own notes mentioned that the two families were more interested in putting their dispute to rest than in establishing the exact amount of land, as they considered the property to be of "minor value." The two families recorded an agreement to be bound by Anderson's "survey" in 1894, establishing that the George family owned north of the agreed upon line and the Sewer family owned south of it, even while acknowledging the survey to be imperfect but satisfactory due to the time and cost that would be involved in a proper survey in comparison to the value of the land.

---

[1]*Newfound Management Corporation v. Sewer,* 885 F. Supp. 727, United States District Court for the District of the Virgin Islands, Division of St. Thomas and St. John, 1995.

Another agreement in 1913 to further divide lands on each side of the line included the following statements:

It is true that the above mentioned [acres] are only given as guess and approximately as no measuring ever has been made, and the acreage thus spoken of can consequently be less—or more but the difference can not be so great, that we could own [substantially more]. We could think it possible, that some of the land in the East End of St. Jan never has been entered in the Matricul, as the land was considered worthless and also now partly is considered of hartly any value. . . .

It is understood, that the acreage, mentioned above is only calculatory and approximately and may be found different, when any measuring should be made. There is however no misunderstanding amongst us with regard to the boundaries. [grammatical and spelling irregularities in original] Pl.'s Exs. 16, 173.

It is this final theme—that the parties involved knew where the boundaries were—that later surveyors had to accommodate when resurveying the area, starting in 1956, for the creation of new lots. In unraveling the ultimate mess of approximate descriptions for well-known boundaries, the court reviewed the location and title for each of the disputed tracts and the general practice of surveying before turning its attention to and devoting a lengthy section of its published opinion to "the Virgin Islands and the special problems presented by surveys conducted in rural areas."

To further confound matters, early surveyors appear to have followed regional surveying practices based on custom. Second, the Danish history of the Islands complicates conducting historical research. Reviewing essential reference documents such as deeds and land lists is more intricate since the surveyor must first collect and perhaps translate old documents to trace ownership of property. Moreover, the Virgin Islands system of recording is idiosyncratic and requires some familiarity. Third, a warm climate, heavy vegetation and rough, often hilly, terrain in undeveloped areas slows a surveyor's team, making fieldwork difficult and time-consuming.[2]

In one footnote to this statement, the court noted the need to understand Danish land description and recording practices as one hindrance to understanding land conveyances on St. John. The Danes had:

. . . divided the rural land into large agricultural tracts called 'estates' to grow products such as sugar cane. These estates, each with a distinctive name, were

---

[2]885 F. Supp. 727 at 751.

further subdivided into separate tracts known as "quarters." When land was conveyed, the historical name, referring to a particular estate and quarter, was commonly used as a geographical unit to identify and describe the transferred portions.[3]

Despite the difficulties, even to the point of an "abundance of chaos" as noted in this case in relation to boundary line locations, the charge of the modern surveyor to walk in the same footsteps as the original surveyor is not altered for the sake of convenience. Determining the original intent of parties to a conveyance may require historical and legal research to supplement the more usual undertakings of record research and field reconnaissance. In this instance, knowing that topographical features such as ridges and streams served as monuments and boundary lines in the Virgin Islands helps explain why the Georges and the Sewers knew exactly where their boundaries were despite lack of measurements. Further, knowing that Virgin Island surveyors merely estimated land quantities and disregarded nonarable lands in their calculations explains why reported acreage cannot be relied upon as evidence of a tract's limits. Finally, recognition that those same early surveyors used hardwood posts as boundary monuments means that finding one of these early markers is evidence of having encountered original survey work. Without this background, early surveys and descriptions would be terribly misconstrued.

Perhaps more complicated than merely understanding the method of surveying and describing land is awareness of the laws in effect at the time of an original conveyance. While this is fairly straightforward when deciphering descriptions written in areas under United States jurisdiction, early patents and grants often have origins in European legal systems, and in areas where sovereignty was juggled back and forth between nations, establishing which legal system applied can entail significant historical research. The city of St. Louis, Missouri, provides one such example, having begun as a French outpost in Upper Louisiana in 1764. Spain took possession of St. Louis in 1770 under the Treaty of 1762, but ceded it back to France in 1800, which eventually ceded the entire Louisiana Purchase to the United States in 1803, including St. Louis. So which nation's laws apply to land grants originating between 1764 and 1772, made by the French and Spanish authorities respectively?

The briefest of responses to this question applies to all countries, outside of dictatorships, which is that between those countries ruled

---

[3] Ibid.

by the "law of nations" (rooted in the Roman Empire's legal system governing interactions between foreigners and provincial subjects), "according to which the rights of property are protected, even in the case of a conquered country, and held sacred and inviolable when it is ceded by treaty, with or without any stipulation to such effect; and the laws, whether in writing, or evidenced by the usage and customs of the conquered or ceded country, continue in force till altered by the new sovereign."[4]

Obviously, any reader of early deeds and grants must know this to apply appropriate definitions of terms, which can vary between cultural customs. Even the term *grant* has a specific meaning when related to treaties, incorporating "not only those which are made in form, but also any concession, warrant, order or permission to survey, possess or settle, whether evidenced by writing or parol, or presumed from possession. . . ."[5]

Treaties between parties within our own borders, however, are governed by yet another set of rules, established by the conquerors and without regard for the Native American inhabitants already utilizing the land. A separate set of statutes and regulations exist within the American body of law to address the alienation and allotment of such lands, based on our European roots rather than incorporating any sense of equity for the people residing here before colonists and pioneers. To understand the rights included with or excluded from a patent or grant on lands that has been the subject of a treaty, the reader of the description cannot assume that the same rules of construction apply as in other sectors of the population on this continent. The sequence of treaties with American Indians and later patents issued by the Government Land Office or other distributor of public lands is therefore of prime importance in determining title to the land and the extent of title allowable by laws in effect at the time.

8.2.2 Clarity and Completeness: Extrinsic Evidence

The drafter of a description has a professional responsibility both to the immediate client and to the public affected by that description, whether adjoiners or future owners of the described tract. Therefore, clarity, completeness, and accuracy are crucial elements, creating a document that both expressly articulates the grantor's intent and is legally sufficient to distinguish the property from all others. Part of

---

[4]*Strother v. Lucas,* 37 U.S. 410 at 436, Supreme Court of the United States, 1838.
[5]Ibid.

assuring a description's completeness and therefore its legal sufficiency (meaning that it can be located on the ground without ambiguity) requires inclusion of references to any documents or conditions that affect understanding the description's meaning. Did the parties draw up a sketch, did they plant stakes in the ground, or were they formalizing a prior oral agreement? Did the plan upon which everyone now relies exist at the time of the original conveyance, or was it drawn up later as an after-the-fact explanation of what was meant but never formally stated? As professionals, we cannot insert our own version of the sequence of events leading up to the conveyance, and must strictly adhere to the facts as they actually occurred, even if the outcome is not as clean a resolution as we and our clients would like.

This last situation is the focal point of *Komadina v. Edmonson*,[6] a case quoted and relied on by courts beyond the state in which it was argued, primarily for the following line:

> The grantor's intent must be ascertained from the description contained in the deed which must itself be capable of being reduced to certainty by something extrinsic *to which the deed refers*.[7] *[Emphasis added]*

The Edmonsons claimed that the Komadinas' deed was void because the intended conveyance could not be located or identified sufficiently from the descriptions. Since the Komadinas had never physically occupied the land, they had to rely entirely upon their deeds to support their claim of title rather than claims of possession, adverse or otherwise.

The Komadinas traced their paper title to 1939, when the incorporated Town of Atrisco, New Mexico issued four deeds to members of the Chavez family. One parcel, Tract 331, had been conveyed to Procopio Chavez with the following description:

> A certain tract of land situate in School Dist. No. 28, Bernalillo Co. New Mexico, Bounded on the North by a Road and on the East by land of Doloritas Chavez and on the South by a Road and on the West by the Atrisco Land Grant. Being one of several tracts of land allotted from the Atrisco Land Grant and more particularly described as follows:
> Measure on the North 210 feet
> Measure on the East 1037 feet
> Measure on the South 210 feet
> Measure on the West 1037 feet
> contains five acres of land more or less.
> Tract No. 331

---

[6]468 P.2d 632, Supreme Court of New Mexico, 1970.
[7]468 P.2d 632 at 634.

Deeds for the other three parcels (Tracts No. 328 to Tonita A. de Chavez, No. 329 to Adela Chavez, and No. 330 to Dolorcitas Chavez) were written similarly, with each description referencing one of the other deeded lands as one of its boundaries and each being bounded by unnamed roads on the north and south. Eventually all four tracts came under the ownership of Procopio Chavez, who transferred $7^{1}/_{2}$ acres of the land to the Komadinas.

Responding to the Edmondsons' challenge to their title, the Komadinas presented a 1943 map made on wrapping paper, showing all the tracts divided into lots, but with no ties so that the lots and tracts floated in space with no reference system. Although roads bounded the lots in agreement with the descriptions, no roads existed at the time of the original conveyance.

Procopio Chavez testified that he had accompanied his father when the latter had driven four pipes into the ground (two of which still stood at the time of the lawsuit), but that he didn't know which corners the presently found pipes were intended to mark: were they related to the four 5-acre Chavez tracts, or to the 7.5 acres conveyed to the Komadinas? But as these pipes had apparently been set after execution of the original 1939 deed, none of them marked original intent. And so the extrinsic evidence of called for monuments and roads proved to be insufficient, since the markers had not existed at the time of original conveyance.

The surveyor for the Komadinas testified that he had surveyed the area in which the tracts were situated, locating tracts 328 through 331 from the 1943 wrapping paper plat, which had been provided to him by the Town of Atrisco. The surveyor could not locate the land from the information contained in the deeds themselves, and the deeds referred to no extrinsic information from which the land could be located. As a result, the New Mexico Supreme Court confirmed the deed's lack of sufficiency to identify and therefore to transfer the land to the Komadinas, 30 years after the tract's creation.

How could this situation have been avoided? Obviously, the process of conveyancing is more formalized now than it was even in the middle of the 20th century, much less a century before then. But it is still possible to overlook important details that will affect future readers of our descriptions. In *Komadina v. Edmonson*, the plan that provided the key to understanding the words of the descriptions was a very informal drawing—but the plan's roughness and informality are irrelevant. The crucial point is that it was *never referenced in the deed*.

Based on the mention of roads in the description that then appeared only on the wrapping paper drawing and not on the ground, it seems clear that the description's author had relied upon the plan. Any reference to that drawing, even if only to mention the date and for whom it was prepared, would at least have allowed the Komadinas to present the plan as a viable candidate matching the reference. The court would then have had to verify its authenticity and admissibility, but the door would be opened to this line of arguments.

We can and should, of course, be complete in our references to any evidence we wish included in our descriptions. One approach is to identify the plan's title, the surveyor who signed it, the plan's original and latest revision dates, and any other distinguishing features. We can also specifically mention that a reduced copy of the plan is attached to and made part of the description. While this copy may be too small to read without a magnifying glass, it at least provides evidence of the plan's existence and intent to make it part of the description.

As an extra step in professional responsibility, appropriate recordation of the plan (whether in the Hall of Records, County Surveyor's office, or other official repository) assists in preserving the evidence for the future. And then the recordation information (where, when, and other recordation identification) should be included in the description as well.

8.2.3 Clarity and Completeness: Consider the Future

The description is an integral part of the contract known as a deed, and so any clauses that indicate some future change to the boundaries or interests must be crystal clear and beyond multiple interpretations. When those wishing to suggest the presence of ambiguities in a deed to further their own interpretation of the document can do so, inevitably they will. When the owners of a Wisconsin tourist cabin business decided to sell part of their land along with the structures and improvements, the deed they prepared read as follows:

A parcel of land in the North half of Section 16, T 13 N, R 6 R, Sauk County, Wisconsin.

Commencing at the Northwest corner of the Northeast quarter (NE $^1/_4$) Section 16, thence East along the Northline [sic] of said Section 16, 241.5 feet, thence South 66 feet to the point of beginning of this description, thence South 789 feet, thence East 334 feet, thence North 88 degrees 8 minutes West 460.75 feet to the centerline of Highway "12-13," thence Northeast along the centerline of said Hwy. "12-13" to a point 66 feet South of the North line Section 16, thence East along a line parallel to the North line of said Section 16, 655 feet to the point of beginning.

Except that portion lying South from cottages commonly known as cottage #2 and cottage #4, the excluded area being about 120 feet by 460 feet; excluded in this conveyance are cottages commonly known as cottage #1, cottage #3, and cottage H-F H-R;

The right to insert the correct legal description is hereby reserved and the parties hereto agree to insert the correct legal description immediately after the premises are surveyed. To identify the portion excluded by this conveyance attached Exhibit "A" is made part of this contract.[8]

Although a sketch was attached to the deed, showing the proposed line demarcating the excluded area, the purchasers closed on the property in 1961 without a survey. Three surveys were subsequently made in 1961, 1962, and 1963, but the sellers chose to refer to none of them in the description. After a number of arguments over the right to use the "excluded" land, the sellers sued to eject the buyers in 1975, after the buyers had already made their final payments under the contract.

The Wisconsin Supreme Court affirmed the trial court's finding that the buyers' interpretation and survey reasonably identified the land, so that the deed had not been in violation of the Statute of Frauds. The buyers' surveyor's work had been executed in accordance with standard surveying practice, and approximated the directions of the written description as closely as possible. He created a roughly rectangular exception, which did not take into account cottages #1 and #3 because they did not exist at the time of his survey. This, of course, was the basis of the sellers' complaints that there was not enough room to walk around those cabins built after the survey's completion.

The sellers had never acted on the rights reserved to them by the deed language, leaving the buyers in apparent legal limbo. What can such a case teach us? While we prefer to think the best of others, it often is not in our best interests to trust their future behavior. Much can happen in even a short span of time to change minds and introduce new variables never intended in the original transaction. Instead, the description should set definite limits to any such reserved future rights in terms of time, conditions, and extent of those rights. After all, the deed is a contract, and the description is the body of the agreement between buyer and seller. Definiteness ultimately protects both parties and minimizes litigation.

8.2.4 Clarity and Completeness: Addressing Three Dimensions

Subsurface conveyances have been with us since the origins of mining, whether for gold, coal, or selenium, but only since the 20th century

---

[8] *Zapuchlak v. Hucal*, 262 N.W.2d 514 at 515–516, Supreme Court of Wisconsin, 1978.

has the issue of air rights become a source of contention and litigation. The advent of air travel meant that certain easements were presumed over earthbound holdings in order to allow first hot air balloons and then airplanes to fly over them. These easements are for the most part still unwritten and merely presumed, except in the form of flight path easements allowing planes to approach and depart airports with specific glide path heights, widths, and shapes set by Federal Aviation Administration requirements based upon the type of aircraft utilizing a particular facility. These easements do, however, affect the use of the ground beneath them, limiting heights of structures and requiring safety features such as lights or markers to alert pilots to the presence of such structures.

But we now encounter other air rights, involving ownership rather than mere easement rights. Condominiums present one early example of the need to be able to describe three-dimensional spaces, not always bounded by walls. These present a special challenge to the description writer, who must understand not only the physical dimensions of the space but also its reference to the elevation of the ground. Beyond condominiums, there are other three-dimensional parcels of interests that do not touch the ground, as we witness the construction of "bridge" structures connecting upper floors of tall buildings and overhanging others below. Definitely, this tests the skills of both writer and reader of descriptions of such parcels, not a task for the geospatially challenged.

Condominium ownership of a unit includes joint ownership with other owners, and the common spaces through which a unit owner must pass to access his or her own unit are owned not 100 percent but in some fixed percentage of ownership. The division of ownership extends to shared facilities such as recreation centers and utility rooms housing the electrical services entering the structure, facilities that make the individual units both habitable and desirable. The description must specify these widely varying full and shared interests in a manner that informs the purchaser of all the rights and restrictions associated with the condominium and provides notice to all who seek to use the easements included in the condominium.

The situation of shared ownership of some elements and singular ownership of others can present unanticipated obstacles to those wishing to use easements within a condominium—the deed and its elements must be thoroughly understood before assuming where rights exist. For example, a Virginia case arose in the 1990s when Media General Cable asked the Sequoyah condominium owners' association for permission to install its cable wires in "compatible easements" within the common areas of the development in order to fulfill the requests of various condominium unit owners.[9]

While there is more to the case than will be related here, three of the four possible easements suggested by Media General Cable were clearly not available for its use due to the specific purposes and entities for which the easements had been created and to which the cable company had no relationship. The fourth area was a blanket utility easement established by the condominium's master deed. However, installing Media General Cable's facilities based on that master deed without express permission from the condominium owners' association would have effectively resulted in a taking of private rights. This is because all interests in the easement were private, due to nature of condominium ownership. It also provides the answer as to why the master deed stipulated that poles and other utility equipment could be attached to residential units only after securing permission from the Condominium Owners' Council.

Related to the question of the extent of rights within a condominium even when there is permission to share a common easement, *Multi-Channel TV Cable Company v. Charlottesville Quality Cable Corporation*[10] began as an argument over the sharing of an easement when one utility removed the facilities of the other in order to provide its own service. However, the case provides other insight as to the extent of condominiums.

Again omitting the nasty fighting details of the case, ultimately the court determined that both the language creating licensed rights in the easements and the maps of the condominium limited the rights of utility facilities to run *to* the individually owned units, but not *into* them. The utility companies had no right to access the interiors of the buildings and units with their wiring through the easements without specific permission from the owners' association (to penetrate the walls) and the unit owners (to enter the inside of the units).

Thus, in reading deeds for condominiums, we cannot assume extension of rights in easements to seemingly qualified users, and the full details of the master deed and unit deeds must be fully understood. Completeness and clarity once again must guide the writing of these documents to assure no trampling of private or shared ownership rights.

---

[9]*Media General Cable of Fairfax, Inc. v. Sequoyah Condominium Council of Co-Owners,* 737 F. Supp. 903, U.S. District Court for the Eastern District of Virginia, Alexandria Division, 1990, and 991 F.2d 1169, U.S. Court of Appeals for the Fourth Circuit, 1993.
[10]65 F.3d 1113, U.S. Court of Appeals for the Fourth Circuit, 1995.

Familiarity with statutes in the state where the condominium resides is provides the description writer with appropriate legal context for such ownership rights within that state. While similar in many respects, laws do vary between jurisdictions, some spelling out the specifics of what is considered part of the condominium unit and distinguishing what is considered part of the common area. One may explicitly identify the right to pass through common areas as a nonexclusive appurtenant easement. Another may define such rights differently. The description writer must choose words carefully to comply with both legal and private interests.

Furthermore, what is legally acceptable in a given state may not provide clarity to the purchasers of a condominium unit and its related interests in common areas. Statutes in many states provide that a legally sufficient description merely needs to refer to the recorded declaration of condominium, also sometimes called deed of declaration or master deed (any of which legally creates the condominium), and bylaws (which provide the definitions of units, air space, common and limited common elements, etc.), thereby including all appurtenances to the unit whether or not separately described, including but not limited to the specified undivided share of common elements.

But statutes also acknowledge that the allocated percentages of ownership in common elements can change if the condominium association approves addition or combination of units, and that boundaries between adjoining units may be relocated. Such conditions require amendments to the Declaration; however, someone recycling a description from a prior conveyance for areas maintaining the same unit identification but bearing new limits prevents discovery of any changes. Imagine the following possibilities in which reference only to the original Declaration and "Unit 5B" can present misleading information:

- Unit 5B has merged with Unit 6A to create a new larger space
- Unit 5B has acquired garage rights not mentioned in the original declaration
- Unit 5B has been subdivided into new Units 5B and 5B1

Statutes specific to the state in which the condominium is located often provide formats and language for condominium descriptions, but the distinction between conventional condominiums and horizontal or land condominiums must be understood. A unit in a conventional condominium consists of a unit of space described by horizontal and perimeter boundaries, while horizontal or land condominiums units are only described by the perimeter boundaries shown on the recorded condominium plan. In this latter instance, the units are treated the same as subdivision lots on a recorded plat, except that the common elements (streets, etc.) are privately owned by the unit owners.

Often, the surveyor will be required to prepare a certification reading somewhat as follows:

> The construction of the improvements is substantially complete so that the material, together with the provisions of the declaration describing the condominium property, is an accurate representation of the location and dimensions of the improvements and so that the identification, location and dimensions of the common elements and of each unit can be determined from these materials.

Such a statement equates to an assertion that the development was built in accordance with the plans, or essentially so, and that the unit can therefore be identified by reference to the master deed, declaration of condominium, and other legal documents. (We recommend that surveyors carefully consider the language of such a certification, and modify it to report substantial conformance with the legal documents rather than making assertions of absolute accuracy.) This is not, however, the description of the conveyed interests, which can take several forms. The first example is of the briefest form suitable for either a conventional or horizontal regime condominium. A plan or map of the condominium is attached to and made part of the declaration, so that it is of public record and automatically made part of any description referring to that declaration.

> Unit 1201 of "Knight's Bridge Commons", a Condominium according to the Declaration of Condominium thereof, recorded in Official Records Book 4528, Page 92 of the public records of Brevard County, Florida, together with its undivided share in the common elements.

The second example can be modified to be suitable for any three-dimensional area, condominium, or otherwise.

> Three dimensional, subterranean part of vacation of a portion of Broad Street for the purpose of extending and maintaining existing subway station facilities at the intersection of Broad and Spruce Streets in the City of Philadelphia, Pennsylvania.

> That portion of the following described part of Broad Street lying below a horizontal plane having an elevation of 45.2 feet (North American Vertical Datum of 1988) and lying above a horizontal plane having an elevation of 27.2 feet, described as follows.

> (Horizontal metes and bounds follow.)

An ancient maxim states "to whomsoever the soil belongs, he owns also to the sky and to the depths" (or rather, in Latin: *"cujus est solum ejus est usque ad coelum et ad inferos"*[11]). Such private rights in Europe were generally subject only to fishing rights (in favor of the public) and mineral rights (in favor of the sovereign). Current property rights battles often address air rights, including the space over existing structures on the ground, with the landmark case of *Penn Central Transportation Co. v. New York City*[12] opening the door to arguments that use of air space belonging to the owner of a plot of land on the earth's surface can constitute a taking of privately held rights for which just compensation must be paid, based on protection offered by the Fifth and Fourteenth Amendments to the Constitution. Existing mineral rights, rights to "ancient lights," viewsheds, air glide paths for airports, and other three dimensional rights connected to a parcel should be considered in creating or construing the written description. Conditions and restrictions affecting grantor and grantee should be specified (and fully read and agreed to by both parties to a deed), as in the following paragraphs from a pipeline easement:

> Grantor does further grant and convey to Grantee the right, privilege, and authority to trim, cut, and remove WITHOUT NOTICE to Grantor, such tree branches, roots, shrubs, plants, trees and vegetation which might, within the exclusive discretion and sole judgment of the Grantee, interfere or threaten the safe, proper or convenient use, maintenance or operation of said gas facilities within the Easement Area.

> Grantor shall have the right to use, occupy, and enjoy the surface and air space above the easement area for any purpose which does not, within the exclusive discretion and sole judgment of Grantee, interfere or threaten the safe, proper or convenient use, occupancy or enjoyment of same by Grantee. Grantor agrees, however, that no buildings or structures shall be erected over said facilities of Grantee.

8.3   REGIONAL LEXICON AND LOCAL PRACTICE

A description written in one part of the country and read by someone from another part of the United States may find that terminology is either obscure or confusing. Some words are not used at all in some areas, while other terms are used more universally but have completely different meanings. Take, for example, the buttonwood tree. In Florida,

---

[11]*Black's Law Dictionary*, 1979.
[12]438 U.S. 104, United States Supreme Court, 1978.

this is *Conocarpus erectus*, a tropical and shrubby mangrove tree that thrives primarily along the southern coast of Florida's peninsula. Elsewhere, "buttonwood" is just one of the common names for the more widely spread *Platanus occidentalis*, also known as sycamore, American sycamore, American plane tree, and occidental plane tree.

Readers who are required to apply the words to the ground must understand local lexicon and regional terminology. A completely erroneous interpretation of location and rights will be the alternative. The case of *United States of America v. Certain Land Located in the County of Barnstable, Commonwealth of Massachusetts*[13] illustrates how the legal system addresses unusual language.

The United States had exercised its powers of condemnation to conform to the Cape Cod National Seashore Act (16 USC §459b to §459b-8), a federal law meant to preserve the seashore. Of the 251 acres involved in this suit, Beede claimed ownership of all, and therefore all the related compensation. But Bessay claimed 8 acres of this land. In a trial without a jury, the prior court had decided in favor of Beede. The contentious clause that Bessay brought up on appeal for reevaluation and definition was "hollow of the beach so called." Did this term approximate the seashore, as Beede's expert testified, or was it a valley several hundred feet back from the shoreline, where Bessay's expert placed it?

Beede's expert, who testified he had never seen this phrase before, answered this question as follows:

> It is generally felt, or it is generally conceded, from a conveyancing point of view, that the term "beach" includes the land which is between high and low water mark; therefore, it seems to me that the term "hollow of the beach" probably refers to some monument which occurs between high and low water marks. So, in fact, even if the description does not carry all the way down to the low water mark, it certainly includes to some monument—whatever it may be—that is, something below high water mark, which would, therefore, be presumably under water during high tide.[14]

In contrast, Bessay's expert described "the hollow of the beach" as a dry valley, which the lower court had visited and therefore described as lying between two large dunes. The Court of Appeals, in overturning the prior decision, chastised that decision:

> ... it was improper procedure to interpret a phrase of six words by defining one of them (beach), making a guess as to another two (hollow of), and omitting the

---

[13]674 F.2d 90, U.S. Court of Appeals, First Circuit, 1982.
[14]674 F.2d 90 at 91–92.

last two altogether. The full phrase was "hollow of the beach so called." Every word, presumptively, has meaning. The most apparent meaning of 'so called' is to suggest something special, or idiosyncratic. It would be an appropriate label, for example, if the word "beach" were being used with other than the court's normal meaning. As to what was suggested by "hollow," the Beedes' witness, instead of, for all that appeared, drawing a probable meaning out of the air, might better have consulted the dictionary. "Hollow," inter alia, means "a low spot surrounded by elevations; a small valley." Webster's New Int. Dict. (3d ed.). [Bessay's expert], precisely. [Beede's expert's] opinion was valueless.[15]

Beede's expert failed to note the words *so called* in his testimony, a term that "should have invited inquiry, and consideration of the use of this phrase in earlier deeds."[16] He had also noted that the description was grossly in error compared to the location of Bessay's property. Unconvinced by his testimony, the U.S. Court of Appeals applied the entire six-word phrase to the site conditions and the deeds in question to find in favor of Bessay.

Historic and local units of measurement may also prove confusing to the uninitiated. While linear perches and rods are generally understood to be 16.5 feet in length (note "generally": such units may possibly differ due to national origins of measurement in a locale), area referred to in square perches and roods may be less easily recognized.

A *square perch* is one perch by one perch, or the equivalent of the square rod, presuming the 1607 English standard of a perch or rod being one-fourth of a Gunter's chain in length. One hundred sixty square perches correspond to an acre of land surface. A *rood* is equivalent to one-quarter of an acre, or 40 square rods/perches. The interrelationship between these units means that it is not all that uncommon to find land descriptions reciting distances combining rods, chains, perches, and acres.

Beginning at a stone corner of David Sheppard's land by the edge of the drained meadow: thence North eighty one degrees and a half East, thirty perches and four links to a ditch; thence by the same, North twenty degrees West, sixty perches and thirteen links; thence South fifty five degrees and a half West thirteen perches and eighteen links; thence South sixteen degrees and a half West, twelve perches; thence South eight degrees and a half East, forty two perches to the beginning corner. Containing eight acres and thirty two square perches of drained meadow, more or less.

---

[15] 674 F.2d 90 at 92.
[16] 674 F.2d 90 at 93.

Further afield from the original 13 colonies, a mixture of units arose with non-British settlers and the resulting grants from absentee sovereigns. Both French and Spanish land grants occurred in various parts of the United States, even alternating in some of the same locales. As we have mentioned, both French and Spanish governors had provided land grants to early settlers in the area of St. Louis, Missouri before the Louisiana Purchase in 1803, and French, Spanish, and English authorities had issued grants to settlers in present-day Mississippi for a century before that area came under the control of the recently formed United States. Thus, not only different units of measure intermingled (such as the Spanish *vara* and French *arpent,* the latter being a term describing both length and of area, and differing between North America and France), but also the cultural differences in how land was divided and shaped. French "long lots" and Spanish "ranchos" were not completely in conformance with the later rectangular survey system of the public lands.

Further complicating matters was the use of arpent land division by both the French and the Spanish in what was to become part of the Louisiana Purchase, with the French version generally 2 to 4 arpents wide along the riverbank by 40 arpents deep and the Spanish version more often 6 to 8 arpents wide but of the same depth as the French. Confirming preexisting rights was not always an easy proposition. Litigation did not always resolve these problems, particularly as some units were not standard even in their home countries, adding to the confusion about locating land rights. For example, variations in the vara, depending on the locale, could mean the unit measured anywhere from 33 to 43 inches. Definitely, the local standards in place at the time of original patents and surveys are a necessary part of recovering ancient lines.

Contemporaneous local practice of surveying is another strong factor in understanding descriptions. While modern surveyors are used to measuring horizontal distances with the push of a button, the process of surveying with a chain introduced a number of physical factors, both relating to the terrain and to the chain itself. In hot weather, the chain would expand, meaning that a measurement of one chain stretched in the summer's heat would report a physical distance shorter than a measurement of one chain in the cold of winter if a correction was not factored into the recorded distance; a long chain reports a distance as being shorter than it actually is, and a short chain reports the distance as being longer than its true physical extent. It was common practice in many locales to "flat chain," or measurement along the earth's surface, rather than the more difficult horizontal process (called

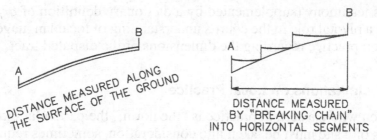

FIGURE 8.1   In the sketch on the left, the surveyor is measuring along the surface of the ground with the chain following the contours of the ground. The surveyor on the right is measuring between surface features with the chain supported at the ends and as level as possible.

"breaking chain"), requiring many short and careful measurements that had to be added up for a cumulative line length. Figure 8.1 compares the two approaches to measuring the earth's surface.

Local methodology leads to local terminology, as in the case of *George M. Murrell Planting & Manufacturing Company v. Dennis*,[17] in which Murrell took title to a Louisiana property with the following description:

> A lot of land containing two superficial arpents bounded on the West by land of Peter Egerton, North by Jos. Henry, and East by land formerly of P.O. Hebert, together with all improvements.

While Murrell could trace his unbroken chain of title back to 1797, the Dennis title originated with a 1920 partition among the heirs of the same Egerton named in Murrell's deed, and this ultimately defeated the Dennis claim. However, for purposes of the current discussion about local practices, we will focus on the reference to "two superficial arpents" in Murrell's deed, and what this means. The court listened to testimony from Murrell's surveyor, Hargrave, who stated that there is a difference between a "linear arpent" and a "superficial arpent":

> Basically, a "superficial arpent" is the surface area of property, the length and width of which are equal and consist of one linear arpent. Hargrave said the property sold to the Dennises was not deep enough to allow two superficial arpents. So they had to make it one and one-half superficial arpents.[18]

---

[17]970 So. 2d 1075, Court of Appeals of Louisiana, First Circuit, 2007.
[18]970 So. 3d 1075 at 1084.

This testimony (supplemented by a dictionary definition of *arpent*) played a pivotal role in the court's understanding of local language and common practice regarding the dimensions of the disputed tract.

8.3.1 Limitations on Local Practice

But even where a certain practice is "the norm," the particular circumstances of a site must be taken into consideration, sometimes requiring the reader of a deed to actually visit the property and always requiring the application of common sense. In the steep North Carolina countryside, the parties in *A. M. Stack v. N. M. Pepper*[19] argued over the boundaries of land that one had conveyed to the other, with one claiming that the distance along the perpendicular surface of the cliff was part of the measured length of the line (as in "flat" or surface measurement) and the other arguing that the original patent had been laid out by horizontal chaining. The difference between these two approaches would mean that certain property (including its timber and minerals) was either included or excluded from the deeded transaction.

Early surveys of patents from the State generally employed surface rather than horizontal measurements, but the North Carolina Supreme Court noted in its brief but pointed opinion that such a presumption as to applying this practice to every survey:

> ... prevails only where it appears feasible and reasonable to have pursued that course. On the contrary, the courts will not assume that the surveyor and chain-bearers procured ladders and climbed over a rugged boulder or cliff situated as in this instance, but that they adopted practicable methods.[20]

8.3.2 Marketable and Registered Title

The concept of marketability is that a purchaser who is well informed about all matters both legal and factual is free from reasonable doubts and therefore willing and ready to pursue and complete a contractual transaction. Relating to real property, marketability means that the title to the land is free from liens or other encumbrances and there is no doubt as to validity of the title.

The process of determining if a property is marketable requires investigation into a wide variety of documents, not just deeds but also mortgages, pending litigation (*lis pendens*), probated wills, creditors'

---

[19]25 S.E. 961, Supreme Court of North Carolina, 1896.
[20]25 S.E. 961 at 962.

liens, and a host of other encumbrances. The process can be complex and difficult, and some states have enacted various registration systems to try to simplify it and clear the record of ambiguities.

The Torrens title system was developed in the 1850s by Robert Torrens to assist him in managing land title registry in his role as Registrar for South Australia. Drawing from several other registry methods, presumably including those employed in insuring shipping, the Torrens system was meant to provide a simple means of examining and registering land titles by creating a single register for each land holding that recorded information about all interests relating to that parcel.

There is a difference between a title registry, as created by the Torrens system, and a deed registry. The latter is a file of documents relevant to the title of land that must be reviewed in order to determine title; a deed registration system provides constructive notice[21] that documents are on record. The Torrens Land Registration System, however, results in a certificate of title issued by the government, providing notice (and guarantee) of *title*—something very different from merely having a deed on record. Torrens is a statutorily created system meant to simplify the transfer of land and provide certainty in proving title of ownership in that land, regardless of any flaws in the chain of title preceding the certificate of title and subject only to specific exceptions established in the statutes creating the system within a given state. The Torrens certificate of title is a single document in which ownership and all mortgages, judgments, and liens are recorded, eliminating the need to do a further records search.

Due to the expense of maintaining the system, many of the states where it once was enacted have since repealed their Torrens Acts, but Torrens and other similarly named land registration systems still exist in nine states as of this writing.

All jurisdictions within the United States must rely on the court system to settle legal disputes about land boundaries and real property interests. In most instances, this entails going before a judge who may or may not have much—or any—experience in real property law, having graduated from law school with the usual generalized education, but perhaps then practicing in specialties such as child custody litigation, corporate law, or criminal defense, but never having so much as read

---

[21]*Constructive notice* provides a means for someone to be aware of something, as opposed to *actual notice,* which provides the actual knowledge. Thus, a recorded deed provides constructive notice of owning land interests to anyone who cares to search the records, while living in a house presents actual notice that someone possesses (but may not necessarily own) the property.

a deed. For this reason, not every decision regarding land and land interests coming from even the upper courts is well founded.

Two of the states employing land registration systems have established special land courts to which such suits are directed. These are not to be confused with early land courts in this country, such as the Virginia Land Court Commission that served to grant lands from the public domain to soldiers after the Revolutionary War, or the United States Court of Private Land Claims (1891–1904), a federal court created to settle land claims in the southwestern territories and states that had been guaranteed to those already living on the land at the time of the 1848 Treaty of Guadalupe Hidalgo. The latter court resolved a very small percentage of the cases brought before it, as it faced a "foreign" system of measurement (the *vara*, which could vary in length depending on the date of the land grant written in such units) and use of unfamiliar terms describing physical limits of land interests.

At opposite ends of the United States, a land court in Hawaii and one in Massachusetts verify land title and descriptions within those jurisdictions. Both are legislated creations, with Hawaii's land court established in 1903 and Massachusetts's in 1898. Both employ the services of title examiners and surveyors in evaluating and then adjudicating ownership and boundary disputes, thereby assuring more consistency and propriety of outcome than in a general court system where the luck of the draw can produce a judge or panel of judges to whom real property law is a foreign language. Proper recordation of a land court's certificate or decree of title effectively guarantees title to registered land, meaning that any prior records in the property's chain of title must yield to the current adjudicated documents. The guarantee embodied in the certificate of title is subject only to state and federal legal exceptions, and is evidence of title against the whole world (*in rem*).

A similar method of "cleaning up the deed mess" is for states to enact Marketable Title Acts instead of Land Registration Acts, allowing title searchers to search back not to the sovereign or other root of a chain of title but instead to some statutorily established "reasonable" period, commonly 20 to 40 years but up to 60 years in a few jurisdictions. This system is based on an owner's having a clear (no conflicting claims, no ambiguities) record chain of title back to its root (patent, deed, court decision, etc.) for that period of time, so that it is marketable; the title is presumably free of all interests recorded prior to the root of the title.

Marketable title acts do not provide for certificates of title, as do land registration acts; instead, they extinguish any claims not filed within a legislated time from adoption and enactment of the enabling legislation. An initial statutory grace period allows owners of real property interests to record or re-record documents that serve as proof of those interests after passage of a marketable title act, but then these interests must be re-recorded every cycle of the statutory period. Claims to rights or interests that have not been renewed or pursued for longer than the act's statutory period are considered lapsed. Exceptions do exist, but only as established by the specific marketable title act passed by the legislature of a given state. Mineral rights in particular are one of the contentious areas often excepted from these acts, so that documents recorded prior to the enactment of the act may sometimes survive.

Incomplete or confusing references preserved for eternity? Or simplified and one-stop record searches? Whether it is a land registration act or a marketable title act in place, only the most complete, clear, and concise description will protect real property interests and stand as strong evidence when reviewed by a title examiner in establishing the paramount rights of title and interest holders.

8.3.3 The Effect of Legislation and Courts

While we live in a country of "united" states, government and court activities do establish differences in practices that are otherwise uniform. Particularly in the early days of forming this nation, land acquisitions were not conducted consistently, due to cultural differences and biases. While deeds or cessions might have been exchanged between various countries vying for ownership of land in the New World, interactions with the Native Americans already on that land more often took the form of treaties drafted by the Europeans and marked as agreed to by the local inhabitants. The new United States' increasing desire for land eventually led to legislated reserved lands to which the Native Americans were expected to move, freeing up more area for new settlers.

Chapter 25 of our federal statutes contains the laws passed by Congress to dictate where the Native Americans were to allowed to live as "sovereign nations" (modified at various times as pressure for westward expansion and desire for mineral and oil rights grew) and what interests in real property individuals can or cannot hold and exercise in those reservations. Nowhere else have private rights been so constrained, but these laws do mean that real property is a different commodity in such areas.

Past court actions are often look to as guidance in establishing land rights or interpreting descriptions of those rights. However, this is only appropriate under certain circumstances. The first consideration must be that the facts of the current situation match or are comparable to the court case one wishes to present in supporting arguments. The relationship between the parties, the root of title for each side, the presence or lack of physical or record or oral evidence regarding boundaries: all of these must be similar enough for the present court to accept a fact scenario's similarity and the relevance of current arguments.

Even when a fact scenario aligns with that of a cited case, the kind of court issuing that prior decision must also be considered. As mentioned before, we have decisions based in law and decisions in equity. Decisions issued by a court of law are based on existing laws and interpretation of their meanings and applicability. Originally, the English monarch established both courts of common law and chancery courts shortly after the Norman Conquest in 1066, with chancery courts meant to rule where no law existed to address a matter. The king's chancellor would also consider cases in which the law courts might rule too rigidly, and this generally meant real property cases were heard in chancery. In a modern chancery court, or court of equity, cases are presented before a single judge, who issues a ruling based on both fact and law; no jury is involved. The judge may have considered unusual factors in formulating a decision, a process of "balancing the equities" as opposed to determining which law applies and applying that law stringently.

The difference between these two approaches to legal dilemmas means that not every outcome can be relied upon as establishing precedent that should guide later decisions. Because real property disputes were one of the first and primary classes of cases assigned to equity rather than law courts, decisions in those matters should be read carefully to formulate arguments but not relied upon as definitive proof relating to the current factual situation.

This being said, the courts do make decisions about boundaries on a regular basis, affecting our understanding of surveying practices and description interpretation. The 1969 case of *Trustees of the Internal Improvement Fund of Florida v. Wetstone*[22] provides such a contrast. At the center of the dispute was a meander line and whether or not it constituted the boundary line between the swamp and overflowed lands and the lands of the state.

We know from the *Manual of Surveying Instructions* that meander lines in the Public Lands Survey System (PLSS) are not property lines. The most current language of the *Manual* addressing this topic reiterates the purpose of meanders since initiation of the PLSS:

---

[22]222 So. 2d 10, Supreme Court of Florida, 1969.

MEANDERING 3-159. The traverse that approximates the margin of a permanent natural body of water, e.g., the bank of a stream, lake, or tidewater, is termed a meander line. Numerous decisions in the United States Supreme Court assert the principle that, in original surveys, meander lines are run, *not as boundaries of the parcel*, but (1) for the purposes of ascertaining the quantity of land remaining after segregation of the bed of the water body from the adjoining upland, (2) for defining the sinuousities of the water body for platting purposes, and (3) for closing the survey to allow for acreage calculations. The ordinary high water mark (OHWM), or line of mean high tide (line of MHT) of the stream, or other body of water, and not the meander line as actually run on the ground, is the actual boundary.[23] *[Emphasis added]*

The 1973 edition of the *Manual of Instructions* was equally clear about distinguishing meanders from boundaries, using very similar language in section 3-115, as does section 226 of the 1947 *Manual of Instructions*. The 1855 version of the *Manual* merely addressed how to meander rather than defining the distinction from boundary lines, but it was issued at a time when the surveying process was still under way and very little of the vast domain had been settled or claimed for there to be much dispute over government-surveyed lines.

The *Trustees* case is rooted in a survey by II. Jenkins in 1875 for the United States government in which he meandered the outer limits of Little Pine Island. After this survey, Florida was granted a patent covering the island under the federal Swamp Lands Act, and Wetstone's predecessors in title took title through a patent and subsequent conveyances in which the description was by government lots.

Wetstone had a bulkhead line established and then applied to the State to purchase all the sovereignty lands between that line and his land. It was at this point that the argument erupted over the location of the line dividing Wetstone's upland interests from the State's interests. Wetstone hired a surveyor, who testified that the flatness of the land, combined with soft spots and hard spots that would affect the placement of a surveying rod, meant that a surveyed line of the mean high tide from the nearest tidal gauge station eight miles away could vary in location anywhere from several hundred feet to a quarter of a mile when it reached Little Pine Island.

Jenkins's 1875 meander line could be located through his field notes and section lines he surveyed on the island. The original survey showed area within the government lots, sections, and fractional

---

[23] *Manual of Surveying Instructions for the Survey of the Public Lands of the United States*, p. 81, United States Department of the Interior, Bureau of Land Management, Cadastral Survey, Denver, CO, Government Printing Office, 2009.

sections patented to Florida, and the state had issued its patent to Wetstone's predecessor using the same description, reciting the same area. Wetstone's surveyor referred to a United States Coast Guard survey of 1866–1867, testifying to no change in the edge of vegetation between 1866 and the litigation a century later. Acknowledging the usual distinction, the Florida Supreme Court confirmed the trial and appellate courts determinations that there could be situations in which the meander line of an official survey could constitute a boundary, such as for areas in low marsh or lacking shore to navigable waters. Wetstone's situation fell into this category, and the meander line did form the boundary between the two claims.

As a side note and in reference to the hierarchy of evidence, here the acreage of Wetstone's claim was the same as had been included in the original patent to his predecessor from the state, acreage that had been based on the area on Jenkins's survey. The Florida court applied rules of construction to uphold quantity of land when ambiguity and uncertainty had eroded the value of other elements in the description that are usually accorded superior worth.

8.4 INTRODUCING UNIFORM LANGUAGE

One of the intentions of this book is to encourage clearer descriptions, and in part that can be accomplished by using language uniformly rather than employing colloquialisms or esoteric terms understood by only a few readers. Plain language with the simplest meanings that still fully express the intent is the best approach to preparing a description of real property interests.

Readers and writers of descriptions should be careful in presuming there is only one meaning to an everyday term or phrase. For instance, while we may all think we know what and where the bank of a watercourse is, the legal definition provides a very specific location, rather than a slope or general vicinity. A number of court decisions have defined the "bank" of a river, but the Supreme Court of the United States provides a very clear example in the case of *Howard v. Ingersoll*.[24] This argument over the boundary along the Chattahoochee River between Georgia and what is now Alabama (formerly territory owned by the United States of America) traces jurisdictional questions back to 1789 through acquisitions of various territories by the United States and

---

[24] 54 U.S. 381, Supreme Court of the United States, December 1851 Term, decided May 27, 1852.

subsequent formation of states. The Judiciary Act of 1789 amounted to a cession of lands between the United States and Georgia, establishing what was to become the boundary between Alabama and Georgia as:

> West of a line beginning on the western bank of the Chattahoochee River, where the same crosses the boundary-line between the United States and Spain, running up the said River Chattahoochee, and along the western bank thereof, to the great bend thereof, next above the place where a certain creek or river called "Uchee," (being the first considerable stream on the western side, above the Cussetas and Coweta towns,) empties into the said Chattahoochee River; thence ...

After discussing low water marks and the characteristics of a river's bed, the highest court in our country made the following statement, the essence of which still guides all boundaries coinciding with the bank of a watercourse:

> When banks of rivers were spoken of, those boundaries were meant which contain their waters at their highest flow, and in that condition they make what is called the bed of the river. ... Such a line may be found upon every river, from its source to its mouth. It requires no scientific exploration to find or mark it out. The eye traces it in going either up or down a river, in any stage of water.[25]

Seventy years later, Oklahoma and Texas brought their ongoing battle over division line between them to the highest court in the nation five times.[26] As part of its decree, the Supreme Court of the United States once again defined *bank* in the process of establishing the location of the south bank of the South Fork of the Red River where it intersected the 100th meridian. The Court quoted the report of the commissioners appointed to run, locate, and mark the boundary between the two states along the Big Bend Area:

> 5. The south bank of the river is the water-washed and relatively permanent elevation or acclivity, commonly called a cut bank, along the southerly side of the river which separates its bed from the adjacent upland, whether valley or hill, and usually serves to confine the waters within the bed and to preserve the course of the river.

---

[25] 54 U.S. 381 at 415–416.

[26] 256 U.S. 70, Supreme Court of the United States, 1921; 265 U.S. 500, Supreme Court of the United States, 1924; 267 U.S. 452, Supreme Court of the United States, 1925; 272 U.S. 21, Supreme Court of the United States, 1926; 273 U.S. 93, Supreme Court of the United States, 1927.

6. The boundary between the two States is on and along that bank at the mean level attained by the waters of the river when they reach and wash the bank without overflowing it.

7. At exceptional places where there is no well defined cut bank, but only a gradual incline from the sand bed of the river to the upland, the boundary is a line over such incline conforming to the mean level of the waters when at other places in that vicinity they reach and wash the cut bank without over-flowing it.[27]

In reiterating the same court's opinion in *Howard v. Ingersoll* 70 years earlier, this 1924 report makes clear that the public at large does not necessarily understand words commonly used in describing boundaries. Description writers should strive to use terminology based on uniform understanding, and if the situation is technical or subject to misunderstanding, include adequate information in language that a layperson can grasp.

Aside from the preceding illustrations of the different conceptions of *bank,* riparian boundaries are often litigated to determine whether it is the high water mark, low water mark, center of the overall water body, or center of the deepest channel (*thalweg*) that serves as the dividing line between two claims of real property interests. If there are multiple branches of a stream or river, the description demands specificity regarding which fork or tributary is meant as the line of demarcation.

8.4.1 "Commencing" versus "Beginning"

When describing a tract of land, tying that parcel to a solid reference point is not always as simple as "Beginning at this well-known point." Instead, a reliable, readily identified and easily reconstructed reference may be some distance away from where we want to begin our description. In such cases, we insert a description of the relationship between where we want to begin our description and where our reference point is. This may be a single line or multiple courses, but the opening words of our description must make clear whether we are situated at the place where our description starts to run around the perimeter of the tract or if instead we are at the distant reference point.

The point where the body of our description starts circumnavigating our tract is the point of beginning. This point may be marked on the ground or established as a corner of an abutting adjoining parcel. But

----

[27]265 U.S. 500.

if we need to tie our point of beginning to another reference point, perhaps the intersection of two street lines or the southerly boundary of a recorded plat where it intersects another line of ownership, that reference point is the point of commencement. This description of this particular point may be similar in nature to the point of beginning, in that it can be a physical or record reference. But the distinction between the beginning and commencing points must be made crystal clear because the public at large does not understand the difference between what they represent. In section 6.4.2 we mentioned litigation based on a mistaken belief that the point of commencement was where the litigant's property actually began, rather than a reference point to guide the reader toward the point of beginning. People who understand that references to physical markers generally carry more weight than references to the amount of land being conveyed can be swept up into arguing for monuments that do not mark their own boundaries.

In section 6.4.2 of this text we also offer suggestions on how to verbally and visually distinguish between commencing and beginning points in a written description. The courts sometimes undertake the establishment of these two points, and may have a different approach from what we offer. For example, the disputed common boundary of Grant Parish and Rapides Parish in Louisiana, as settled in 2006 litigation, reads as follows:

> Commence at a point located at latitude 31 degrees 23 minutes 29.0203 seconds north, longitude 92 degrees 37 minutes 55.3123 seconds west, said point marking the point of beginning. From point of beginning, run northeasterly to the eastern terminus, said terminus having a latitude of 31 degrees 28 minutes 46.9037 seconds north, longitude 92 degrees 11 minutes 52.1355 seconds west. The line described hereinabove is a grid line based upon the Louisiana State Plane Coordinate system, North Zone. The geographic coordinates of the end points of the line are based upon the North American Datum of 1983.[28]

As a side note, reporting longitude and latitude in this format requires more decimal places for the seconds than those familiar only with bearings expect to read. But this level of precision is necessary due to the rounding errors in converting between degree and coordinate values of latitude and longitude. See section 6.4.3.4 of this text to read more about significant figures.

---

[28]*Rapides Parish Police Jury v. Grant Parish Police Jury*, 924 So. 2d 357 at 363–364, Court of Appeal of Louisiana, Third Circuit, 2006.

8.4.2 Word Choices, Grammar, and Punctuation

Throughout this text we have provided examples of proper use of words and phrases. In many cases these are terms used in everyday language that seem ordinary, but in fact have very specific legal meanings. As a result, an overwhelming number of lawsuits have revolved over the definition of a single word or phrase, the outcome altering the interpretation of a tract's described boundaries. In section 6.6.1 we discussed careful and proper use of language, including both technical and nontechnical words. Proper grammar is both a matter of professional presentation and a source of confusion. And as we mentioned in section 6.6, punctuation, including overuse or lack of it, can completely alter the meaning of a description. While the case of *Brockel v. Lewton* (319 N.W. 173, Supreme Court of South Dakota, 1982) did not pivot on punctuation, the placement of a comma played a role in delaying the sellers' performance on the contract to sell their land. From the court report:

> This document described the land, in part, as including the SW $^1/_4$, NW $^1/_4$, and S $^1/_2$ SE $^1/_4$, Section 35, T16N, R14E; ... Subsequently, this description was found to be erroneous because Lewtons do not own the entire NW $^1/_4$ of Section 35. They do, however, own the NW $^1/_4$ of the SE $^1/_4$ of Section 35. The trial court found that the placement of the comma after NW $^1/_4$ in the land description was a clerical error.[29]

For the uninitiated, it is the comma after the term "NW $^1/_4$" that creates the error. As written, this description includes the SW $^1/_4$ of Section 35 AND the NW $^1/_4$ of Section 35 AND the S $^1/_2$ of the SE $^1/_4$ of Section 35. Instead, the description should have conveyed the SW $^1/_4$ of Section 35 along with the NW $^1/_4$ and S $^1/_2$ of the SE $^1/_4$ of Section 35. Figure 8.2 illustrates the distinction.

8.5 BREAKING OLD HABITS

As human beings, we are creatures of habit. This is sometimes because we are comfortable with the patterns in our lives, and sometimes simply because it is more time consuming or stressful to do things differently from how we have done them in the past. Other times our own experience is not broad enough to suggest alternative methods.

---

[29]319 N.W.2d 173 at 174.

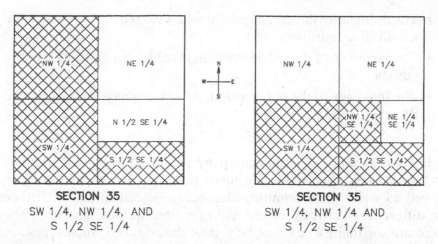

FIGURE 8.2 *Brockel v. Lewton:* **The sketch on the left is the property as described with the additional comma and the sketch on the right is the property as described without the erroneous comma.**

The strongest impetus for defending "the way things have always been done" is simply inertia. But with just a little effort we can improve the written record while protecting our clients' interests (and our own) in the process of preserving evidence of boundaries.

8.5.1 Repeating Old Descriptions Verbatim

Many title companies and attorneys prefer to reuse descriptions from prior transactions on the basis of consistency and not creating new ambiguities. But there is a risk with such verbatim repetition. The following short list of examples caution against such a practice:

- The boundaries have not changed, but the original description was inconsistent with physical conditions on the ground (physical calls in the description do not reflect what is on the ground).
- The boundaries have not changed, but original markers have been disturbed or removed.
- The boundaries have not changed, but new corner markers have been set where none existed previously.
- The boundaries have not changed, but the original description was inconsistent with record conditions at the time of the creation of the real property interests (incorrect references to adjoiners or plats, etc.).

- The boundaries have changed, for whatever reason (consolidation, subdivision, partition, etc.).
- The language of the old description is subject to multiple interpretations.
- The language of the old description uses language no longer readily understood.

In some instances, the title company and/or attorney will accept an updated or new description but insist that the old description be included as well: "Said premises also being described as..." If there are differences between the old and new descriptions (and invariably there are, or there would not be a new description), such a practice introduces ambiguity and conflict into the real property transaction. Which description should be believed? Which (if either) is more accurate? Which most closely conforms to the intent of parties creating the current boundaries?

Returning to the earlier example of one of the author's purchase of a lot that had been conveyed using the same description over a period of 40 years of transactions, the closing attorneys were displeased with the removal of that old description and its replacement by one that altered bearings and added references. The bearing of the line through the party wall was updated so that it actually went through the shared or party wall and not into the neighbor's dining room (based on the newly surveyed angle between the street line and the offending wall). The party wall was also referenced in the description for the first time in 90 years, when the house initially had been constructed and sold. Further, the new description reflected the boundary line agreement and shared driveway easement that had transpired in the intervening 20 years since the immediately prior conveyance of the property. Merely recycling the old description verbatim would have perpetuated error and ambiguities, losing valuable evidence in the process.

8.5.2   Destroying Evidence

While old descriptions may contain errors or ambiguity, they also often contain information that is helpful in later years to those who are attempting to retrace boundaries. Sometimes our efforts to clean up the language of a document will purge the only information that can possibly resurrect an understanding of the conditions and facts surrounding the creation of a parcel.

It is common practice to use well-known local landmarks in descriptions. These vary widely in nature: buildings with which everyone in town was familiar ("the northeast corner of the top step of MacPhee's Pharmacy"); outstanding trees ("the old red oak at the bend of Stackhouse Road, bearing three blazes on the north"); sites of historic importance ("in the center of Freehold Square where the old gallows stood"). But local landmarks can change or disappear over time, a fact unanticipated by the authors of descriptions. For this reason, multiple connections between past and present are helpful to provide clues to ancient evidence. From a 1923 deed:

> Beginning at a monument set on the westerly edge of a public road leading from Brookville (formerly Millville) to Waretown on a course north 40 degrees 32 minutes west 32-$\frac{1}{2}$ feet from the center thereof and in line of land formerly of William Griffith (year 1851) and running thence south east along said Griffith land, (1) south 40 degrees 32 minutes east 1304.5 feet to an iron pipe in the line of what was formerly known as "The Little Mill Property" but now a part of this tract; thence northeast along said Mill tract . . . (7) north 51 degrees 21 minutes west 1471.8 feet to a monument on the northerly edge of the so-called Jones Road and corner to land formerly of Charles Penn, see deed dated May 3, 1861, Book of Deeds 26, page 95, later of Ann Eliza Penn, see deed dated May 13, 1873, Book 71, Page 201. . . .

While this particular description was probably recycled in a number of deeds before being recorded in 1923 (and in fact was pressed into service again in 1955), it provides a significant amount of history in terms of record ownership, local property names, and local road names. It also provides insight into the fact that some of the older tract lines have disappeared as new property configurations evolved. The extent of the descriptive narrative may seem overwrought in an age when everyone seeks brevity. However, the context and resources it provides are invaluable.

8.5.3 Jargon, Colloquialisms, and Abbreviations

Description writers are like many other technical writers, often falling into the presumption that the world at large shares a common understanding of certain terms that in reality are familiar only to a select group of specialists. For this reason, to reach the largest audience we need to choose words carefully and spell out words rather than using abbreviations. At the same time, we have to avoid the other extreme of using colloquialisms, informal language that is often imprecise, when more technical terminology is more appropriate.

As an example of the difficulty with abbreviations, some of the readers of this text may immediately recognize the letters *ACSM* as designating *American Congress on Surveying and Mapping*, a professional organization of surveyors, geodesists, cartographers, and geographic information systems specialists. However, the health organization *American College of Sports Medicine* shares the same initials and same abbreviation. Different readers will interpret abbreviations differently, based on their own experiences. As further illustration, one of this book's authors had left a trade publication on the kitchen table and found her research physician spouse frowning while trying to decipher *BMP*. The title of the article both on the cover (where space is at a premium) and within the feature article (where more space is more generously available) used only that same abbreviation. It wasn't until the second paragraph of the article that the magazine author managed to write out a translation for my baffled spouse, whose own personal frame of reference yielded "bone morphogenic protein," completely unrelated to the article's topic of "best management practices" for stormwater detention.

The common use of abbreviations on a survey for features found in the field is often unaccompanied by any legend to translate groups of letters or symbols and squiggly lines for the uninitiated. That is a strike against preserving evidence in and of itself. But the second strike occurs when the written description perpetuates this presumption of common knowledge. *IP* has multiple meanings even in surveying parlance: iron pipe, iron pin, iron post. This means that surveyors will not know specifically what kind of a marker was found or set in the field. Readers with disparate backgrounds may guess none of the above and substitute some other unintended words. Spelled-out words should accompany the first use of any abbreviation in a description—or any other document meant for public use.

But sometimes we slip into a writing style that is too informal, too conversational, too filled with idioms, slang, jargon, and clichés when we mean to simplify the presentation. While the goal of "plain" language is a worthy one, use of unexplained local or colloquial phrases does not further the cause and can instead confuse matters further.

The term *stob* is one such villain. Some surveyors in the southeastern United States may understand this to be a wooden stake or post, although in more recent times the term is used to refer to a reinforced metal bar also known as *rebar*, the latter also sometimes erroneously expanded into "reinforced iron bar" even though it is made of steel. Geographical outsiders probably never will have heard of a "stob" before encountering the term in a description. If such foreigners are new landowners who want to walk the perimeter of their newly purchased tract, they will have no idea what to look for, particularly as the word does not have a single meaning.

8.5.4 Sentence Construction and Punctuation

In pursuit of full explanation, we easily fall prey to the lure of long sentences. Each term needs a definition; each bit of physical evidence demands explanation of size, material, and relationship to lines and corners; each reference to a bounding adjoiner requires a recitation of recorded pedigree. Combined with long-windedness, improper or missing punctuation can completely obliterate the intended message.

The abuse of punctuation falls into three main categories: too much punctuation, too little punctuation, and use of the wrong punctuation. While we will not dive into a lecture on writing, we offer a few observations to improve the clarity and meaning of land descriptions.

In the category of "overused punctuation," probably the comma is at the top of the list. We saw in section 8.4.2 how an extra comma can alter the meaning of a phrase or sentence. But there is also the matter of making it easy for the reader to follow the meaning of a sentence with appropriate punctuation. Some writers seem to believe that a comma is required every five or so words to break up a long sentence, rather than actually breaking a long sentence into two or more shorter ones. The result is choppy and hard to follow. The point of the sentence becomes lost among the various phrases. Related information should be grouped together, separated by punctuation if in list format.

At the other extreme are the description writers who use only periods and practically no other form of punctuation. In section 6.6 we saw an example of an ambiguity introduced by lack of a single punctuation mark. Commas, semicolons, and colons are important tools in assuring that only one meaning can be deduced from a sentence. For those who are unsure of the proper use of these punctuation tools, many tutorials are available on the Internet.

Long sentences often should be broken into separate sentences. Each new thought warrants its own new sentence; stream of consciousness is not a good way to write a land description. Proper punctuation can keep a sentence focused when the words run beyond two lines of text.

Misused punctuation marks can degrade both the meaning and the quality of a document. Professionalism suffers when a comma is pressed into service when a semicolon is called for or an apostrophe is thrown into a word where it doesn't belong. Spell checking with a word processor does not always catch such errors, and the best results come from actually reading the document before signing it and sending it out the door.

8.5.5 Copying a Writing Style

Too often, we read recent descriptions that are difficult to follow because of format or language. Many of us learned to write descriptions following the style of more experienced colleagues in the offices where we worked. But just because the old-timers before us used the phrase "to wit" at the end of their opening paragraphs does not mean that we are tied to that same practice. Instead, we should strive for modern and clear language, maintaining flexibility to adapt to the various situations we encounter. Not every parcel will be a simple rectangle at the intersection of two improved public streets with established right-of-way widths. Not every parcel will even be located on the surface of the earth, as some may entail air rights and others may be subterranean. And not every description is for a nice, neat fee simple conveyance without any conditions. We may need to find words to express the intent to convey a determinable one-third interest in a conservation easement over only part of a particular tract of land.

Throughout this text, numerous samples of descriptions have illustrated the points that we have wished to make. Some of these have been fictitious, and some have been pulled from real deeds. Perhaps none of them have been perfect in terms of being concise or perfectly punctuated. But it is our goal to provide enough examples for readers to compare and contrast in establishing their own style while incorporating the best of the best examples and modifying them to suit particular situations as they arise. It is not our intent to provide a "one style fits all" template, but instead to encourage creativity in the pursuit of writing the best possible description. The basic tools to achieve this are knowledge of the technical aspects of description writing, a solid understanding of the intent and purpose of the description, and a good dictionary. An educated and creative mind that enjoys a linguistic challenge is an extra advantage.

AFTERWORD

Land. Language. Law.

Land values change, language evolves, and law reflects shifts in social viewpoints. As a result, we must read descriptions with an understanding of their contemporaneous context. But this also means that as forms of land ownership and use change, language will adapt to describe those changes, and the legal system will shift to accommodate them. For example, condominium forms of ownership arrived on the scene late in the 20th century, requiring entirely new forms of written descriptions and laws to regulate them.

Understanding that the world continues to change, it is our responsibility as readers and writers of land descriptions to keep an open mind to the past and the future. No matter the format, clear and consistent language will always serve to protect the interests of every party to a land transaction.

With these thoughts in mind, we hope our readers will continue to improve the written record by taking time to research the background of existing records, creating clearer land descriptions when new ones are needed, and preserving the evidence by assuring descriptions are publicly recorded.

TABLE OF CASES

INDEX